# West Nile Virus Infection

# West Nile Virus Infection

Editor

**Francisco Llorente**

MDPI • Basel • Beijing • Wuhan • Barcelona • Belgrade • Manchester • Tokyo • Cluj • Tianjin

*Editor*
Francisco Llorente
Centro de Investigación en
Sanidad Animal (CISA-INIA),
CSIC
Spain

*Editorial Office*
MDPI
St. Alban-Anlage 66
4052 Basel, Switzerland

This is a reprint of articles from the Special Issue published online in the open access journal *Pathogens* (ISSN 2076-0817) (available at: https://www.mdpi.com/journal/pathogens/special_issues/West_Nile_Virus_Infection).

For citation purposes, cite each article independently as indicated on the article page online and as indicated below:

LastName, A.A.; LastName, B.B.; LastName, C.C. Article Title. *Journal Name* **Year**, *Volume Number*, Page Range.

**ISBN 978-3-0365-7240-6 (Hbk)**
**ISBN 978-3-0365-7241-3 (PDF)**

# Contents

**About the Editor** . . . . . . . . . . . . . . . . . . . . . . . . . . . . . . . . . . . . . . . . . . . . . . **vii**

**Francisco Llorente**
West Nile Virus Infection
Reprinted from: *Pathogens* **2023**, *12*, 151, doi:10.3390/pathogens12020151 . . . . . . . . . . . . . . 1

**Freude-Marié Bertram, Peter N. Thompson and Marietjie Venter**
Epidemiology and Clinical Presentation of West Nile Virus Infection in Horses in South Africa, 2016–2017
Reprinted from: *Pathogens* **2021**, *10*, 20, doi:10.3390/pathogens10010020 . . . . . . . . . . . . . . . 5

**Gamou Fall, Diawo Diallo, Hadiza Soumaila, El Hadji Ndiaye, Adamou Lagare, Bacary Djilocalisse Sadio, et al.**
First Detection of the West Nile Virus Koutango Lineage in Sandflies in Niger
Reprinted from: *Pathogens* **2021**, *10*, 257, doi:10.3390/pathogens10030257 . . . . . . . . . . . . . . 21

**Cécile Beck, Isabelle Leparc Goffart, Florian Franke, Gaelle Gonzalez, Marine Dumarest, Steeve Lowenski, et al.**
Contrasted Epidemiological Patterns of West Nile Virus Lineages 1 and 2 Infections in France from 2015 to 2019
Reprinted from: *Pathogens* **2020**, *9*, 908, doi:10.3390/pathogens9110908 . . . . . . . . . . . . . . . 33

**Serena Marchi, Emanuele Montomoli, Simonetta Viviani, Simone Giannecchini, Maria A. Stincarelli, Gianvito Lanave, et al.**
West Nile Virus Seroprevalence in the Italian Tuscany Region from 2016 to 2019
Reprinted from: *Pathogens* **2021**, *10*, 844, doi:10.3390/pathogens10070844 . . . . . . . . . . . . . . 49

**Hana Zelená, Jana Kleinerová, SilvieŠikutová, Petra Straková, Hana Kocourková, Roman Stebel, et al.**
First Autochthonous West Nile Lineage 2 and Usutu Virus Infections in Humans, July to October 2018, Czech Republic
Reprinted from: *Pathogens* **2021**, *10*, 651, doi:10.3390/pathogens10060651 . . . . . . . . . . . . . . 59

**Érica Azevedo Costa, Marta Giovanetti, Lilian Silva Catenacci, Vagner Fonseca, Flávia Figueira Aburjaile, Flávia L. L. Chalhoub, et al.**
West Nile Virus in Brazil
Reprinted from: *Pathogens* **2021**, *10*, 896, doi:10.3390/pathogens10070896 . . . . . . . . . . . . . . 71

**Virginia Gamino, Elisa Pérez-Ramírez, Ana Valeria Gutiérrez-Guzmán, Elena Sotelo, Francisco Llorente, Miguel Ángel Jiménez-Clavero, et al.**
Pathogenesis of Two Western Mediterranean West Nile Virus Lineage 1 Isolates in Experimentally Infected Red-Legged Partridges (*Alectoris rufa*)
Reprinted from: *Pathogens* **2021**, *10*, 748, doi:10.3390/pathogens10060748 . . . . . . . . . . . . . . 85

**Elisa Pérez-Ramírez, Cristina Cano-Gómez, Francisco Llorente, Ani Vodica, Ljubiša Veljović, Natela Toklikishvilli, et al.**
Evaluation of West Nile Virus Diagnostic Capacities in Veterinary Laboratories of the Mediterranean and Black Sea Regions
Reprinted from: *Pathogens* **2020**, *9*, 1038, doi:10.3390/pathogens9121038 . . . . . . . . . . . . . . . 101

**Juan-Carlos Saiz**
Animal and Human Vaccines against West Nile Virus
Reprinted from: *Pathogens* **2020**, *9*, 1073, doi:10.3390/pathogens9121073 . . . . . . . . . . . . . . . 121

# About the Editor

**Francisco Llorente**

Francisco Llorente, Ph.D, is a Principal Investigator at the Animal Health Research Center (Centro de Investigación en Sanidad Animal) (CISA-INIA), CSIC. His research focuses on arboviral diseases of animals, mainly in the areas of pathogenesis, diagnosis and surveillance. He is currently involved in different national and international projects in relation with arboviral zoonosis, including those caused by flaviviruses. He finished his BS degree in Biology in 1992 at the Universidad Complutense de Madrid (Spain) and received his PhD from the same University in 2000, working at the National Institute for Agricultural and Food Research and Technology (Instituto Nacional de Investigaciones Agrarias, INIA). Then, he moved to the Universidad Politécnica de Madrid for a postdoctoral term for five years and since 2006, he has been working at CISA-INIA as a virologist and scientific researcher.

*pathogens*

*Editorial*

# West Nile Virus Infection

**Francisco Llorente**

Centro de Investigación en Sanidad Animal (CISA-INIA), CSIC, Valdeolmos, 28130 Madrid, Spain; dgracia@inia.csic.es

West Nile virus (WNV) is a mosquito-borne pathogen that belongs to the *Flavivirus* genus (family *Flaviviridae*). The virus is maintained in nature in a rural cycle between mosquito vectors, mainly *Culex* species, and avian hosts. Spillover from this cycle occasionally results in outbreaks in horses and humans where, in severe cases, the infection can induce neurological signs such as meningitis and encephalitis, and in some cases can lead to death. The virus has spread in the world in this century, and an increase in the number of outbreaks as well as their severity has been observed in Europe in recent years. Considering all of these aspects, a multidisciplinary approach including public, animal, and environmental health, is the best option to increase the knowledge on this problem, making this disease an excellent example of a "One Health" issue.

In this Special Issue of *Pathogens*, readers can find contributions covering topics on different aspects of the virus, including its pathogenesis, vaccines, diagnosis and epidemiology, in both humans and animals. We have eight contributions in the form of original research and a review.

Six of the research papers published in this Special Issue are focused on epidemiological studies on different areas of the world including different continents where the epidemiological history of WNV has been completely different. In Africa, the virus has been present since its first description in 1937 in the West Nile district in Uganda and can now be considered as endemic in the continent. In Europe, WNV emerged in the 1960s, but only sporadic cases occurred in the continent up to the 1990s. Since then, the frequency and relevance of outbreaks have increased, reaching the highest number of cases in 2018 (2083 reported cases in the European Union). The presence of the virus in the Western Hemisphere is much more recent. The virus was detected for the first time in New York in 1999 and quickly spread across North America causing an epidemic with a high number of human cases and bird deaths. Between 2001 and 2004 it was present in the Caribbean, and in South America viral genome was detected in Argentina in 2006, Colombia in 2012 and Brazil in 2018.

Epidemiology in South Africa is the focus of the article from Bertram et al. [1] which describes the epidemiological situation and clinical presentation in horses during 2016–2017. Samples from horses that were positive to the disease were obtained by passive surveillance and compared with non-infected horses. The results obtained indicated that WNV cases were higher in young and non-vaccinated animals, but were also correlated with the season, high altitude, and the breed of the animals. The increase in cases observed in 2017 can be attributed to environmental factors favouring vector population. Data obtained indicate that annual vaccination is especially recommendable for animals younger than five years-old and highly purebred breeds, and that the best season for vaccine application is spring. A second study carried out in Africa in this Special Issue describes the detection of Koutango virus in sandflies in Niger for the first time [2]. Koutango is a divergent genetic variant that, depending on the authors, can be considered a different lineage of WNV, being the most distant lineage showing high genetic distance in relation to others, or a related virus species. Koutango has only been detected in Africa. Unlike other WNV lineages, it has mainly been found in rodents and ticks, and shows a higher virulence in mice than other

**Citation:** Llorente, F. West Nile Virus Infection. *Pathogens* **2023**, *12*, 151. https://doi.org/10.3390/pathogens12020151

Received: 9 January 2023
Accepted: 13 January 2023
Published: 17 January 2023

previously assayed WNV strains. The virus detected in sandflies was isolated, sequenced, and inoculated in mice, causing 100% mortality. High pathogenicity observed in mice and a possible higher range of vectors as indicates the detection in sandflies, suggest that Koutango can be a relevant emerging pathogen.

Three other contributions reflect the epidemiological situation in recent years in different European regions. Beck et al. [3] report outbreaks in Southern France between 2015 and 2019 in humans, equids and wild birds. The isolation of WNV lineage 1 in 2015 in the Camargue region suggests an endemic situation in the area in the year with the most important outbreak in horses. In 2008, lineage 2 emerged producing the most important epidemics affecting humans in France. The increase of cases in the Camargue area in 2018 is correlated with an increase in *Culex pipiens* population. Additionally, this study also describes the presence of the virus in areas without previous WNV circulation with infections in humans, horses and wild birds. Additionally, the isolation of WNV lineage 2 from a raptor in 2018 demonstrated the emergence of lineage 2. A second original paper that focus in Europe describes a serosurveillance [4] in humans in the province of Siena in Italy from 2016 to 2019, in an area without human cases since 2017. Serum samples (1800) were analyzed by ELISA test, immunofluorescence assay, and virus-neutralization. Although a low prevalence, under 1%, was detected, active circulation was confirmed by the presence of IgM, showing, for the first time the circulation of the virus in the area. The authors indicate a trend inWNV expansion in Central Italy, and although no human cases have been detected in the area the application of preventive measures and surveillance is recommended. A third contribution related to the circulation of WNV in Europe presents a description of human clinical cases in the Czech Republic in 2018 [5]. Clinical, epidemiological and laboratory findings of five patients are described in detail. Four of the patients were confirmed for WNV infection and at least one of these was caused by lineage 2. The fifth patient seems to have been infected by Usutu virus (USUV), showing an antibodies titer against this virus that was much higher than that observed against WNV. The data indicates the simultaneous circulation of WNV lineage 2 and USUV in the country in 2018.

An example of the epidemiological situation in a South American country is presented by Costa et al. [6], showing new genetic evidence of WNV circulation in Brazil after the first genome detection in the country in 2018. Viral genome was detected and sequenced from three symptomatic horses from different Brazilian states. A phylogenetic analysis indicated that independent introduction events from North America were produced for the strains detected in this work and those previously described in 2018. The authors also summarize the past evidence of WNV circulation in Brazil in humans, horses and avian species, including serological detection and provide a climate-informed assessment of the transmission risk of WNV across the country. The previous and new data obtained indicate that both sporadic and endemic local transmission possibly explain the WNV epidemiology in the country.

In relation to WNV pathogenesis Gamino et al. [7] delved into the pathogenesis of WNV in birds. In a previous study the authors demonstrated the susceptibility of red-legged partridge (*Alectoris rufa*) to two Mediterranean WNV strains from Morocco and Spain via experimental infection, showing higher mortality and morbidity in the birds infected with the Moroccan isolate. In this study, differences in pathogenesis have been analysed to explain the different courses of infection and mortality. Although the virus was present in brain and showed a similar viral load, a more acute inflammatory reaction and necrosis were observed for the Moroccan strain. These data suggest that differences in neurovirulence between strains can be more significant than neuroinvasiveness in birds mortality, and that a higher virulence can be caused by a more acute and severe encephalitis.

In the area of diagnosis, Pérez-Ramírez et al. [8] present an evaluation of the capacity of veterinary labs in the Mediterranean and Black Sea regions for themolecular and serological detection of WNV by an external quality assessment (EQA). The study was performed in the context of a European Union funded project (MediLabSecure) whose objective is to create a framework to promote arbovirus surveillance under a One Health perspective. Before the EQA, a training program for WNV detection including workshops for molecular and serological diagnoses was implemented, and the learning as well as acquired capacity of the participants to incorporate the techniques into their own laboratories were determined in this study. Seventeen veterinary laboratories from 17 countries were evaluated. Differential molecular detection by real-time RT-PCR for WNV lineages 1 and 2 and USUV was highly satisfactory, mainly for WNV lineages while less than 50% of laboratories gave correct results (100%) by conventional RT-PCR for generic detection of flaviviruses. In relation to serological detection, results were excellent for the generic detection of WNV antibodies by competition ELISA, but some laboratories failed in the detection of IgM detection in samples with low titers. The evaluation carried out demonstrated that the training program was useful in upgrading the diagnostic capacities in veterinary laboratories of EU-neighboring countries.

In the last manuscript in this collection, Saiz [9] discusses in a review the situation of WNV vaccines in terms of their use in horses, birds and humans. The different barriers encountered to their development and possible commercialization are summarized, considering the fact that although different vaccines are commercially available for horses, none of them have been licensed for humans.

I hope that this Special Issue will contribute to further knowledge on WNV infection. As the Collection Editor, I would like to thank all of the authors, reviewers, and editorial personnel who have made this Issue a reality.

**Conflicts of Interest:** The author declares no conflict of interest.

## References

1. Bertram, F.M.; Thompson, P.N.; Venter, M. Epidemiology and Clinical Presentation of West Nile Virus Infection in Horses in South Africa, 2016–2017. *Pathogens* **2021**, *10*, 20. [CrossRef] [PubMed]
2. Fall, G.; Diallo, D.; Soumaila, H.; Ndiaye, E.H.; Lagare, A.; Sadio, B.D.; Ndione, M.H.D.; Wiley, M.; Dia, M.; Diop, M.; et al. First Detection of the West Nile Virus Koutango Lineage in Sandflies in Niger. *Pathogens* **2021**, *10*, 257. [CrossRef] [PubMed]
3. Beck, C.; Goffart, I.L.; Franke, F.; Gonzalez, G.; Dumarest, M.; Lowenski, S.; Blanchard, Y.; Lucas, P.; de Lamballerie, X.; Grard, G.; et al. Contrasted Epidemiological Patterns of West Nile Virus Lineages 1 and 2 Infections in France from 2015 to 2019. *Pathogens* **2020**, *9*, 908. [CrossRef] [PubMed]
4. Marchi, S.; Montomoli, E.; Viviani, S.; Giannecchini, S.; Stincarelli, M.A.; Lanave, G.; Camero, M.; Alessio, C.; Coluccio, R.; Trombetta, C.M. West Nile Virus Seroprevalence in the Italian Tuscany Region from 2016 to 2019. *Pathogens* **2021**, *10*, 844. [CrossRef] [PubMed]
5. Zelená, H.; Kleinerová, J.; Šikutová, S.; Straková, P.; Kocourková, H.; Stebel, R.; Husa, P.; Husa, P.; Tesařová, E.; Lejdarová, H.; et al. First Autochthonous West Nile Lineage 2 and Usutu Virus Infections in Humans, July to October 2018, Czech Republic. *Pathogens* **2021**, *10*, 651. [CrossRef] [PubMed]
6. Costa, É.A.; Giovanetti, M.; Catenacci, L.S.; Fonseca, V.; Aburjaile, F.F.; Chalhoub, F.L.L.; Xavier, J.; de Melo Iani, F.C.; Vieira, M.A. da C. e. S.; Henriques, D.F.; et al. West Nile Virus in Brazil. *Pathogens* **2021**, *10*, 896. [CrossRef] [PubMed]
7. Gamino, V.; Pérez-Ramírez, E.; Gutiérrez-Guzmán, A.V.; Sotelo, E.; Llorente, F.; Jiménez-Clavero, M.Á.; Höfle, U. Pathogenesis of Two Western Mediterranean West Nile Virus Lineage 1 Isolates in Experimentally Infected Red-Legged Partridges (*Alectoris rufa*). *Pathogens* **2021**, *10*, 748. [CrossRef] [PubMed]
8. Pérez-Ramírez, E.; Cano-Gómez, C.; Llorente, F.; Vodica, A.; Veljović, L.; Toklikishvilli, N.; Sherifi, K.; Sghaier, S.; Omani, A.; Kustura, A.; et al. Evaluation of West Nile Virus Diagnostic Capacities in Veterinary Laboratories of the Mediterranean and Black Sea Regions. *Pathogens* **2020**, *9*, 1038. [CrossRef] [PubMed]
9. Saiz, J.C. Animal and Human Vaccines against West Nile Virus. *Pathogens* **2020**, *9*, 1073. [CrossRef] [PubMed]

pathogens

Article

# Epidemiology and Clinical Presentation of West Nile Virus Infection in Horses in South Africa, 2016–2017

Freude-Marié Bertram [1,†], Peter N. Thompson [1] and Marietjie Venter [2,*]

1 Department of Production Animal Studies, Faculty of Veterinary Science, University of Pretoria, Onderstepoort, Pretoria 0110, South Africa; bertramf@tut.ac.za (F.-M.B.); peter.thompson@up.ac.za (P.N.T.)
2 Centre for Viral Zoonoses, Department of Medical Virology, Faculty of Health Sciences, University of Pretoria, Pretoria 0001, South Africa
* Correspondence: marietjie.venter@up.ac.za; Tel.: +27-12-319-2638
† Current Address: Department of Animal Science, Faculty of Science, Tshwane University of Technology, Pretoria West, Pretoria 0183, South Africa.

**Abstract:** Although West Nile virus (WNV) is endemic to South Africa (RSA), it has only become recognized as a significant cause of neurological disease in humans and horses locally in the past 2 decades, as it emerged globally. This article describes the epidemiological and clinical presentation of WNV in horses across RSA during 2016–2017. In total, 54 WNV-positive cases were identified by passive surveillance in horses with febrile and/or neurological signs at the Centre for Viral Zoonoses, University of Pretoria. They were followed up and compared to 120 randomly selected WNV-negative controls with the same case definition and during the same time period. Of the WNV-positive cases, 52% had fever, 92% displayed neurological signs, and 39% experienced mortality. Cases occurred mostly in WNV-unvaccinated horses <5 years old, during late summer and autumn after heavy rain, in the temperate to warm eastern parts of RSA. WNV-positive cases that had only neurological signs without fever were more likely to die. In the multivariable analysis, the odds of WNV infection were associated with season (late summer), higher altitude, more highly purebred animals, younger age, and failure to vaccinate against WNV. Vaccination is currently the most effective prophylactic measure to reduce WNV morbidity and mortality in horses.

**Keywords:** West Nile virus; horses; South Africa; epidemiology; emerging disease; encephalitis; neurotropic virus; zoonosis

**Citation:** Bertram, F.-M.; Thompson, P.N.; Venter, M. Epidemiology and Clinical Presentation of West Nile Virus Infection in Horses in South Africa, 2016–2017. *Pathogens* **2021**, *10*, 20. https://doi.org/10.3390/pathogens10010020

Received: 30 November 2020
Accepted: 26 December 2020
Published: 30 December 2020



## 1. Introduction

West Nile virus (WNV) is a neurotropic, zoonotic, vector-borne virus in the *Flaviviridae* family [1] and is a member of the Japanese Encephalitis virus sero-complex [2–4]. WNV was first identified in the West Nile District in Uganda in 1937 in a febrile human patient, since which periodic outbreaks were reported in Africa, the Middle East, and Europe [5]. Internationally, human WNV encephalitis was rarely encountered prior to early 1990s [6] but since then, outbreaks of increased severity, from new viral strains, likely of African origin, have occurred in parts of Europe and Asia. Since 1999, the Western Hemisphere was also affected, with substantial WNV disease incidence [7], and WNV has now become a significant globally re-emerging pathogen of importance in international trade [6]. WNV is regarded as the most geographically widely distributed arbovirus with increased incidence and severity of neurological disease in humans and horses as well as high mortality rates in birds in the Western Hemisphere [5], and it is one the leading causes of arboviral encephalitis globally [8,9].

WNV is maintained in nature by cyclic activity in numerous avian and mosquito species. African avian species are thought to be primary, reservoir hosts; they display no apparent signs of infection, which is presumably due to genetic resistance [6,10,11]. Mosquitoes may incidentally spread the virus to humans, horses, and other species, which then act as dead-end hosts due to the lower viraemia achieved in these species [5]. Approximately 20% of WNV infections in horses are symptomatic, with clinical signs ranging from fever to severe neurological signs (90%) and death [6,12]. Case mortality rates in unvaccinated horses range from 30 to 40% [13,14].

Two major lineages of WNV have been identified, lineage 1 and 2, and several less common geographic specific lineages [15–18]. Lineage 1 WNV is predominantly found in the Northern Hemisphere, including Europe, Northern Africa, the Middle East, parts of Asia, and Australia, where a closely related virus, Kunjin, clusters with lineage 1b. Lineage 1 was identified as the cause of deaths in birds, humans, and later horses in New York, U.S.A., in August 1999 [9], from where it spread across Northern America, Canada, and South America [5,15,19]. It is not known how the virus was introduced into the Americas, possibly through the legal or illegal importation of infected birds or the accidental importation of infected mosquitoes by aeroplane [6,9,20]. The WNV strain responsible for the initial North American outbreak was closely related to a WNV strain isolated from a dead goose in Israel during the previous year [19]. WNV-positive cases in horses and humans are reported annually in the U.S.A., with outbreaks differing in magnitude and geographic location, large outbreaks occurring every eight to ten years [21,22].

Lineage 2 WNV is most prevalent in Southern Africa and Madagascar, where it is endemic [2], and it emerged in 2006 in Central Europe from where it has spread to Greece, France, Italy, Germany, and causes frequent outbreaks of neurological infections in humans, horses, and birds [23,24]. Human WNV-positive cases were reported regularly in Europe since 1999, with increasing frequency of seasonal, regional outbreaks occurring in 2012 (935 cases), 2013 (785 cases), and 2018 (1670 cases and 124 deaths). The largest outbreak to date, in 2018, spread across 12 countries in southern and central Europe, and it was attributed to favourable climatic conditions, namely an early spring and very high temperatures during summer [21,25–27]. Although lineage 2 is predominantly associated with neurological infections of WNV in humans and animals in South Africa, a few lineage 1 strains have also been identified, suggesting that migratory birds may also import these strains to the region [28–30]. Passive surveillance of horses with febrile and neurological infections identified WNV cases across the county, particularly in Gauteng, KwaZulu-Natal, the Karoo and the Eastern Cape as well as the Western Cape [12]. Sequence identity amongst South African (RSA) lineage 2 strains indicated an exceptional constancy in the virus, strengthening the suspicion that local circulating foci of the WNV are being maintained in certain areas during the relatively mild, inland plateau winters [10,11]. However, migratory birds may have, on occasion, been the reservoir host responsible for the less common lineage 1 WNV infections [12,31].

Human cases of WNV fever have been consistently diagnosed in South Africa, with the largest outbreak in the Karoo in 1974 [32], followed by an outbreak in 1984 in Gauteng, after periods of unusually high rainfall and flooding in these areas [10]. Approximately 5–15 cases are reported annually by the National Institute for Communicable Diseases (NICD) [11,33]. During 2008–2009, WNV was detected in 3.5% of unsolved cases of human neurological disease in Gauteng provincial hospitals, indicating that WNV is underdiagnosed in human neurological cases [34,35]. This led to a study performed in 2011 to 2012 identifying South African veterinarians as a group with likely similar exposure risk as horses to WNV; 7.9% of veterinarians tested positive for antibodies against WNV, their distribution approximating that of WNV-positive cases detected in animals [36]. The World Organisation for Animal Health (OIE) reported West Nile virus infections in humans in South Africa from 2006 to 2018, with an average of four to ten cases annually. An increase in cases was reported in 2011 (52 cases) and 2012 (36 cases) and only one death was reported in 2014 [37].

The distribution of human outbreaks in South Africa was attributed to the ornithophilic *Culex univittatus* as the main mosquito vector (and to a lesser degree *Culex theileri*, *Culex pipiens*, and *Aedes caballus*) in the Highveld (central plateau) areas [10,38]. Given the right climatic conditions of heavy rains and higher than usual temperatures, *Cx. univittatus* has been responsible for significant WNV outbreaks in humans, despite having a low human feeding rate. Their eggs being very sensitive to desiccation, *Culex* spp. mosquitoes prefer temporary to semi-permanent rain flooded grassland, swamps, or other permanent water collections with emergent vegetation as breeding sites and survive dry winters as quiescent larvae and pupae or dormant adult females [38]. Both the mosquitoes' gonotrophic period and the extrinsic incubation period of WNV in the insect vectors are very temperature dependent and can also be influenced by other environmental factors such as precipitation, hydrology, and humidity [5,39,40], which is why WNV disease tends to occur in late summer or autumn in the temperate, summer rainfall regions.

Research performed in 2000–2001 amongst South African Thoroughbred horses found that 11% of yearlings had already seroconverted against WNV, relative to sera collected approximately 12 months prior [25]. Of their dams, on these widely spread stud farms, 75% had also seroconverted, and yet no neurological clinical signs had been reported in any of these horses [41]. This is consistent with typical WNV occurrence worldwide, as most of the infected horses do not display overt clinical signs (approximately 80%), although viral encephalitis is seen in up to 90% of the symptomatic cases [6,12]. Systematic passive surveillance during 2008–2015 by the Centre for Viral Zoonoses (CVZ), University of Pretoria, confirmed a total of 79 clinical cases of WNV in horses in RSA, of which 91% displayed neuroinvasive disease, with a 34% case fatality rate [12].

Fever, particularly as the main syndrome, is an inconsistent finding in WNV-affected horses, especially when compared to other South African arboviruses such as Sindbis virus (SINV) and Middelburg virus (MIDV) [42]. It seems to be the only clinical sign in equine WNV infection that is not an exclusive reflection of central nervous system (CNS) pathology and may rather be attributed to the horse's immune response to the viral infection. The cytolytic virus' capacity to cause disease depends on its ability to survive in vivo, infect vital cells, and evade immune system recognition, inducing apoptosis in a diverse spectrum of tissues, including neurons [43]. WNV cases developing acute, progressive neuroinvasive disease [14] may show typical encephalomyelitis signs that may range from mild incoordination and weakness to severe ataxia, paresis or paralysis, recumbency, and death [27]. Neurological signs depend on the extent of CNS pathology and may include cranial nerve deficits, as summarised in Table 1 [1,5,6,44]. Up to 40% of recovered horses may show some form of persistent neurological deficit, either gait or behavioural abnormality, post recovery [13,14].

**Table 1.** Typical neurological signs of West Nile virus (WNV) disease in horses as related to damage to three areas of the central nervous system.

| CNS Location of WNV Pathology | Typical Neurological Signs Associated with Specific Area of Pathology |
|---|---|
| Spinal cord pathology | Weakness, ataxia, reluctance to move, paresis, or paralysis affecting one or more limbs, skin or muscle fasciculations, muscle tremors and muscle rigidity. Paralysis of hindlimbs ("dog-sitting"); progressive paralysis of all four limbs usually ending in recumbency. |
| Brain pathology * | Ataxia, dysmetria, hyperaesthesia, and abnormal mentation (ranging from somnolence and depression to agitation and hyperexcitability, even aggression) |
| Cranial nerve deficits | Facial nerve paralysis (VII) including droopy lip, muzzle deviation, or lip twitching. Tongue weakness or paresis (XII), head shaking, head tilt (VIII), and dysphagia (IX, X). Fine tremors of the face (VII) and neck muscles (XI). |

* Pathological changes in medulla oblongata, pons, thalamus, reticular formation, cerebellum and brain cortex.

Currently, no specific treatment is available against WNV infection, but the American Association of Equine Practitioners (AAEP) guidelines recommend supportive treatment and nursing care aimed at reducing the CNS inflammation, preventing self-inflicted trauma, and providing nutrition and oral and intravenous fluid therapy as deemed necessary [5,45].

The control of the disease depends mainly on prophylactic vaccination to stimulate a protective immune response and mosquito management to avoid exposure to infected mosquitoes. Numerous studies have shown that protective immunity against WNV viraemia decreases both the severity of clinical signs as well as the mortality rate [1,5,6,13,45–47]. In RSA, an inactivated WNV vaccine is distributed by Zoetis (Duvaxyn), and a WNV recombinant canarypox virus vaccine is distributed by Merial/Boehringer Ingelheim (Proteq West Nile). These vaccines were licenced after epidemiological studies showed that WNV lineage 2 was associated with fatal neurological disease in horses [2], and a vaccine trial in mice showed that a lineage 1 vaccine cross-protected against lineage 2 WNV infection [48]. The WNV vaccine for horses was widely available in South Africa only as of 2015.

Passive surveillance for arboviruses such as WNV, Wesselsbron (WSLV; *Flaviviridae*), SINV (*Togaviridae*), and MIDV (*Togaviridae*), and Shuni virus (SHUV; *Peribunyaviridae*) has been routinely performed since 2008 for acute febrile and neurological disease in horses and other animals by the Zoonotic Arbo- and Respiratory Virus (ZARV) programme at the CVZ. During 2017, numbers of WNV-positive horses in RSA showed a remarkable increase from those diagnosed, on average, by the CVZ in 2008 to 2016.

The objective of this study was to investigate and describe the epidemiology and clinical case presentation of West Nile disease in horses in RSA from 2016 to 2017. Investigations included measuring the association of fever, acute neurological disease, and death, as well as certain predictor variables, with WNV infection. Predictors included animal demographic factors, vaccination status, environmental factors, and illness or stressful events within 4 weeks prior to sampling.

## 2. Results

A total of 54 WNV-positive cases (6 in 2016, 48 in 2017) were included in the analysis. The age range of WNV-positive horses was 4 months to 18 years (median 5 years). The highest proportion of WNV-positive cases was seen in younger horses (Figure 1), especially those less than 5 years old ($n$ = 30, 55%).

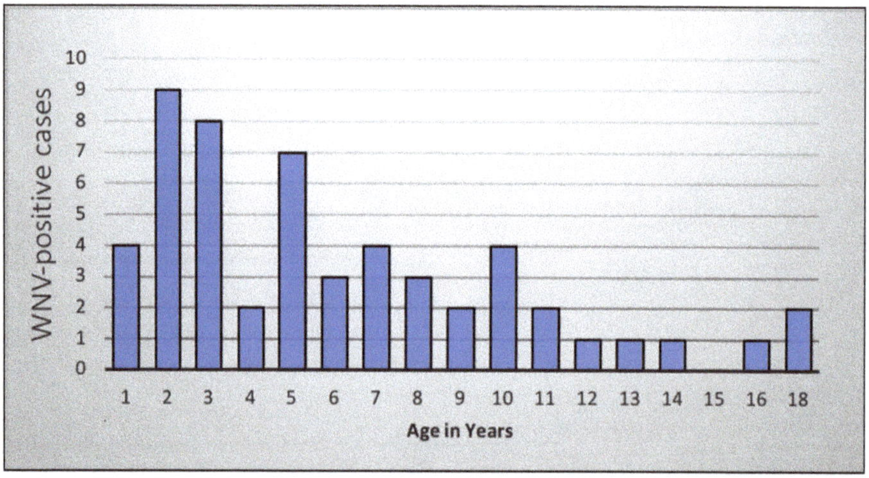


**Figure 1.** West Nile virus-positive cases in horses in South Africa by age in years, 2016–2017.

In total, 15% (8/54) of the WNV-positive cases were co-infected with another virus (four with MIDV and four with equine encephalosis virus, EEV). Of the 120 randomly selected WNV-negative controls, 14 (12%) were MIDV-positive, 15 (12%) were EEV-positive, and 91 (76%) tested negative for all viruses on the ZARV programme testing panel. During the telephonic follow-up, it was determined that most of the controls were not definitively diagnosed. However, various confirmed diagnoses included herpes virus infection, tick-borne diseases (such as babesiosis, Karoo tick paralysis, and vestibular syndrome/facial paralysis from heavy auricular infestations), African horse sickness vaccine reactions, vertebral fracture, guttural pouch mycosis, brain tumours, and severe colic.

Neurological signs, with or without fever (rectal temperature >38.5 °C), were significantly more prevalent in WNV-positive cases than in WNV-negative controls (Table 2). Most of WNV-positive cases in 2016–2017 (48/54, 89%) displayed some neurological signs, of which 54% (26/48) had only neurological signs without fever. Approximately half of the WNV cases (28/54, 52%) had fever with or without neurological signs, fewer than the control group (76/120, 63%), although the difference was not significant (Table 3).

**Table 2.** Disease outcome of West Nile virus-positive cases and WNV-negative controls, by main syndrome, amongst horses detected by passive surveillance of neurological and febrile cases in South Africa, 2016–2017.

| Disease Outcome | | Fever Main Syndrome | Neuro Main Syndrome | Fever and Neuro Main Syndrome | Total |
|---|---|---|---|---|---|
| WNV-positive cases | Deaths | 1 | 14 | 6 | 21 (39%) |
| | Recovered | 5 | 12 | 16 | 33 (61%) |
| | Total | 6 (11%) | 26 (48%) | 22 (41%) | 54 |
| WNV-negative controls | Deaths | 7 | 27 | 9 | 43 (36%) |
| | Recovered | 36 | 17 | 24 | 77 (64%) |
| | Total | 43 (36%) | 44 (37%) | 33 (28%) | 120 |

**Table 3.** The most important clinical signs in the 54 West Nile virus-positive cases and 120 WNV-negative controls, amongst horses detected by passive surveillance of neurological and febrile cases in South Africa, 2016–2017.

| Clinical Signs | WNV-Positive | | WNV-Negative | | Odds Ratio | 95% CI | *p*-Value * |
|---|---|---|---|---|---|---|---|
| | *n* | % | *n* | % | | | |
| Died | 21 | 39% | 43 | 36% | 1.1 | 0.6, 2.3 | 0.824 |
| Euthanised | 16 | 30% | 25 | 21% | 1.6 | 0.7, 3.5 | 0.284 |
| Retained signs post-recovery | 6 | 11% | 3 | 3% | 4.5 | 1.0, 31.0 | 0.053 |
| Neurological signs | 48 | 89% | 77 | 64% | 4.2 | 1.7, 13.7 | 0.001 |
| Ataxia | 40 | 74% | 59 | 49% | 2.9 | 1.4, 6.5 | 0.003 |
| Fever | 28 | 52% | 76 | 63% | 0.6 | 0.3, 1.3 | 0.208 |
| Hindleg paralysis | 19 | 35% | 22 | 18% | 2.4 | 1.1, 5.3 | 0.028 |
| Recumbency | 18 | 33% | 24 | 20% | 2.0 | 0.9, 4.4 | 0.091 |
| Paresis | 16 | 30% | 18 | 15% | 2.4 | 1.0, 5.5 | 0.045 |
| Paralysis | 15 | 28% | 14 | 12% | 2.9 | 1.2, 7.1 | 0.018 |
| Icterus | 11 | 20% | 28 | 23% | 0.9 | 0.3, 1.9 | 0.823 |
| Tremors, fasciculations | 10 | 19% | 8 | 7% | 3.1 | 1.1, 9.9 | 0.041 |
| Foreleg paralysis | 9 | 17% | 4 | 3% | 5.4 | 1.5, 26.8 | 0.008 |
| Anorexia | 8 | 15% | 28 | 23% | 0.6 | 0.2, 1.4 | 0.278 |
| Laminitic stance | 5 | 9% | 1 | 1% | 8.9 | 1.3, - | 0.023 |
| Hyperreactive/Hyperaesthetic | 5 | 9% | 3 | 3% | 3.7 | 0.7, 26.4 | 0.124 |

* Two-tailed Fisher's exact test; significance set at $p < 0.05$.

All types of paralysis (hindleg, foreleg, paresis, and total paralysis) as well as ataxia and tremors/muscle fasciculations were significantly associated with WNV infection (Table 3). Laminitic stance/sensitivity in the feet was noted in 9% ($n$ = 5) of the cases and only one of the controls ($p$ = 0.023). Recumbency tended to be more frequent in cases than controls ($p$ = 0.091). Icterus, anorexia, and hyperreactivity/hyperaesthesia were not significantly associated with WNV infection.

Fatality proportions were similar for both cases (39%) and controls (36%) (Table 2). In both groups, subjects with only neurological signs had the highest fatality proportions (cases 14/26, 54% vs. controls 27/44, 61%), while those with only fever had the fewest fatalities. There was a tendency for WNV-positive cases with fever (with or without neurological signs) to be more likely to recover than those without fever ($p = 0.057$). Of the WNV cases that died, a larger proportion showed hindleg paralysis (9/21, 43%) and total paralysis (8/21, 38%) than foreleg paralysis (2/21, 10%) and tremors (2/21, 10%).

Of the 54 WNV-positive cases, 16 were euthanised due to a poor prognosis, with a median survival time of two days (range 0 to 469 days). Retained neurological signs after recovery were marginally more frequent in cases than in controls (Table 3), which were mainly related to ataxia or neurological instability. Three of these horses, an American Saddler and two Thoroughbred horses, were euthanised at 54–469 days after recovery, ranging in ages: 4 months, 6 years, and 18 years old. The 18-year-old Thoroughbred was also diagnosed with a cardiac tumour at euthanasia. In three other Thoroughbred horses (aged 2.5–3 years), performance was significantly affected by the retention of some degree of clinical signs resulting in early retirement from racing. One of them was sold as a pleasure hack even before racing; another was retired soon after attempting racing (following 8 months' recuperation from WNV infection), and the third had a very unsuccessful racing career.

The breeds mostly represented in WNV-positive cases for 2016–2017 were Thoroughbreds ($n = 26$, 48%), Warmbloods ($n = 9$, 17%), and Arabian horses ($n = 7$, 13%), with the mixed ($n = 3$, 6%) and local breeds (Boerperd and Nooitgedachter) ($n = 2$, 4%) having significantly fewer cases than the purebred horses ($p = 0.009$). Neuroinvasive proportions tended to be similar (90–100%) amongst main breeds, but fatalities appeared to be fewer in Warmblood horses (22%); however, due to small numbers of individual breeds, statistical associations could not be made.

Spatial distribution of WNV cases (Figure 2) showed that most of the equine cases in 2016–2017 occurred in Gauteng ($n = 19$, 35%), KwaZulu-Natal ($n = 14$, 26%), and the Northern Cape ($n = 11$, 20%). The fewest cases were seen in the Western Cape ($n = 5$, 9%), Free State ($n = 3$, 6%), and North West provinces ($n = 2$, 4%), with no reported cases in Mpumalanga, Limpopo, and the Eastern Cape. The largest proportion of cases ($n = 37$, 69%) occurred at the 2nd and 3rd altitude quartiles (1057–1466 m).

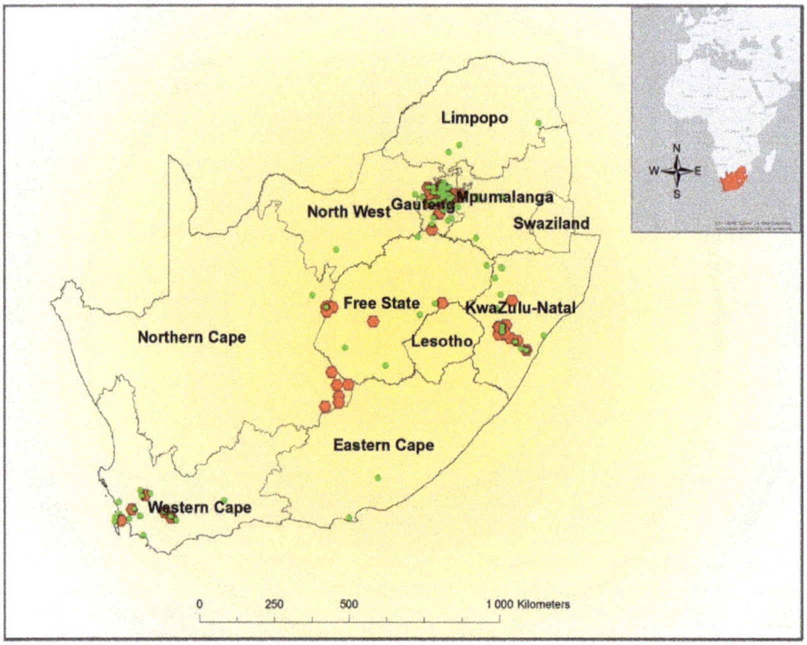


**Figure 2.** Distribution of West Nile virus-positive cases (red) and WNV negative controls (green) in horses used in the study, by province, 2016–2017. Each marker may represent one or more cases at the same location.

In 2016, all six cases occurred during February–April. In 2017, the cases occurred during January–June, with a single case in early December in the winter rainfall region of the Western Cape. March was the month with the highest number of cases in both years ($n$ = 28, 52%) and the greatest proportion of cases occurred during February–April ($n$ = 46, 85%) (Figure 3).

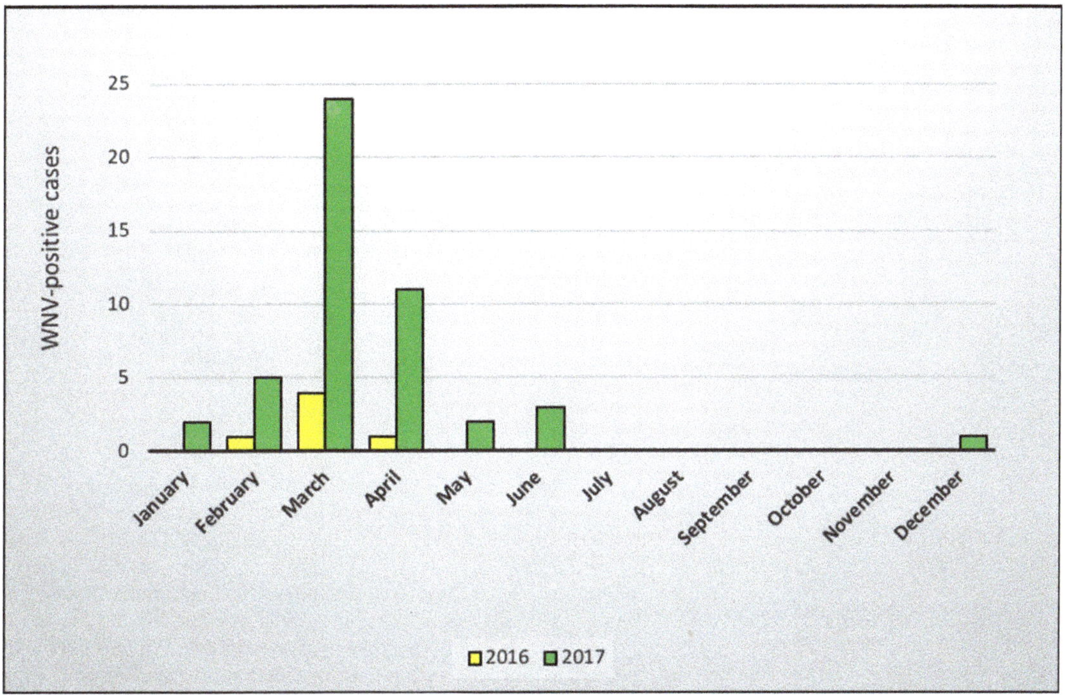


**Figure 3.** Distribution of WNV-positive cases in horses in South Africa by month in which the initial signs were displayed, 2016–2017. The yellow bars show the 2016 cases and green bars show the 2017 cases.

Only 1/54 cases (2%) was reported to have been vaccinated against WNV in the 12 months before sample submission in 2016–2017, in comparison to 9/120 controls (8%). The WNV-positive case that had been vaccinated received an initial WNV vaccination 8 months and a booster vaccination 5 months prior to diagnosis, which were both administered by the owner. Of the 149 herd owners interviewed (cases and controls), only nine (6%) had vaccinated their horses in 2016–2017, whereas 30 (20%) had subsequently vaccinated in 2017–2018.

In the final multiple logistic regression model (Table 4), several variables were significantly associated with WNV infection. WNV cases were more likely to occur during March–April than any other time of the year ($p$ = 0.007), and at an altitude of 1293-1466 m ($p$ = 0.003). The odds of WNV diagnosis were the lowest in mixed and local breeds compared to intermediate and pure breeds, declined with increasing age ($p$ = 0.041), and were lower in WNV-vaccinated horses ($p$ = 0.047). Equine influenza virus (EIV) vaccination, although not significant ($p$ = 0.145), was retained in the model as a confounder, as its removal resulted in substantial changes to the coefficients of the WNV vaccination and age variables. The Hosmer–Lemeshow test indicated the adequate fit of the final model ($p$ = 0.094).

**Table 4.** Final logistic regression model of factors associated with WNV-infection in South African horses with acute febrile or neurological disease detected by the Zoonotic Arbo- and Respiratory Virus (ZARV) programme at the Centre for Viral Zoonoses (CVZ), 2016–2017.

| Variable | Level | Odds Ratio | 95% CI | *p*-Value |
|---|---|---|---|---|
| *Month* | January–February | 5.4 | 0.6, 49.9 | 0.134 |
| | March–April | 18.0 | 2.2, 149.5 | 0.007 |
| | May–June | 4.2 | 0.4, 44.9 | 0.241 |
| | July–December | 1 * | – | – |
| *Altitude* | 16–1056 m | 1 * | – | – |
| | 1057–1292 m | 1.2 | 0.4, 3.9 | 0.807 |
| | 1293–1466 m | 6.0 | 1.9, 19.1 | 0.003 |
| | 1467–1784 m | 1.2 | 0.3, 4.3 | 0.764 |
| *WNV vaccinated* | Yes vs. no | 0.1 | 0.0, 1.0 | 0.047 |
| *Age in years* | Continuous | 0.9 | 0.9, 1.0 | 0.041 |
| *Breed* | Highly Purebred | 3.0 | 0.9, 9.7 | 0.068 |
| | Intermediate Hybrid vigour | 4.9 | 1.3, 18.2 | 0.019 |
| | Mixed and Local | 1 * | – | – |
| *Equine influenza virus vaccinated* | Yes vs. no | 2.1 | 0.8, 5.6 | 0.153 |

* Reference level.

## 3. Discussion

The previous ZARV study of WNV in horses in RSA 2008–2015 [12] reported an average of 10 cases per year. The six WNV-positive cases in 2016 in the current study is consistent with that incidence rate; however, the 48 cases diagnosed in 2017 indicate a marked increase in case numbers. This could be due, in part, to increased awareness of WNV in RSA and sample submissions from suspected cases by owners and veterinarians. Increased awareness of WNV was created over the past few years by pharmaceutical companies' product advertising as well as information disseminated through social media, veterinary congresses, scientific publications, and owner-targeted talks, many of which were facilitated by the ZARV programme, with feedback on WNV-positive cases. It is more likely that the sudden increase in WNV cases detected in RSA in 2017 was largely due to the environmental factors promoting the extensive breeding of WNV vectors, typical of the cyclical nature of these outbreaks, as seen in Europe and the U.S.A. [21]. Ecological and laboratory studies have determined the importance of environmental factors such as temperature, humidity, precipitation, and hydrology in WNV transmission dynamics, enabling mathematical modeling and the forecasting of potential outbreaks [40,49]. Overall, RSA experienced a severe drought in 2015–2016. During the early summer months of 2015/2016, there was very little rain and extremely high temperatures, followed by sudden high rainfall in the late summer months (February–April) of 2016. In particular, the eastern parts of the country had high rainfall in 2017, varying between 75% and 200% of the normal rainfall [50]. Higher rainfall following two dry seasons would have favoured the breeding of mosquitoes. Increased environmental temperatures may, up to a certain threshold, favour replication of the virus in the poikilothermic mosquito vectors, as well as decrease the subsequent length of the extrinsic incubation period and increase the efficiency of transmission of virus to susceptible hosts [40]. The seasonal variation in WNV cases was similar to those described in previous studies, occurring mainly in late summer and early autumn [12,51,52]. The slightly extended period of case distribution may have been attributed to the heavy rains, periodic flooding, and warmer than average temperatures of 2017, which followed the drought and extremely high environmental temperatures (El Niño conditions) of 2015–2016 [53]. In fact, 2017 was reported to be the fourth warmest year in South Africa since 1951, with 2015 and 2016 the two warmest recorded years [53].

The spatial distribution of the 54 WNV-positive cases in horses as detected by the ZARV programme, CVZ, in 2016–2017, followed a similar pattern to those previously detected in earlier years by the ZARV [12]. The largest proportion of cases was detected on the Highveld, mainly in the warm to temperate, summer rainfall zones in the eastern parts of RSA, which was likely due to ideal vector breeding conditions [10,38]. Of the 586 equine test submissions to ZARV in 2016–2017, Gauteng had the highest number of submissions ($n$= 264, 45%) as well as the highest number of WNV-positive cases ($n$ = 19, 35%). Fewer cases were seen in the Western Cape and surrounding areas, which was most likely due to the Mediterranean climate and winter rainfall patterns being less favourable to vector development and not due to a lack of valuable horses or to the underreporting of cases. Only a small proportion of the samples from the Western Cape tested positive for WNV ($n$ = 5/109, 5%), despite the submission of a large number of samples from that region ($n$ = 109, 19% of total sample submissions). Most of the WNV-positive cases that were diagnosed in Thoroughbreds were from stud farms located in KwaZulu-Natal ($n$ = 10), Western Cape ($n$ = 5) and the Northern Cape ($n$ = 4). According to a stud farm survey done by the Thoroughbred Breeders' Association early in 2017 [54], 65% of the Thoroughbred stud horses were located in the Western Cape province, 22% were located in KwaZulu-Natal, and 1% was located in Gauteng. Of those specified on participating stud farms, 52% were classified as youngstock under 3 years old and 45% were classified as adult breeding stock. Only 4% were reported to be non-breeding adult Thoroughbred horses at stud with a negligible number of horses of other breeds (<1%).

Most of the WNV cases were diagnosed using immunoglobulin M (IgM) enzyme-linked immunosorbent assay (ELISA), rather than rtRT-PCR. This is consistent with international findings and is due to the short-lived viraemia [6]. Recent increased exposure resulting in persistent IgM levels in convalescent cases, which may then be incorrectly diagnosed as acute cases, were unlikely, as the WNV-positive cases in this study fit the typical clinical description of acute WNV-positive cases. The total co-infection rate for 2016–2017 approached the 18% level previously reported [12]; however, an increased co-infection rate with both MIDV and EEV was observed during 2016–2017 when compared to previous years [17]. Larger outbreaks of these viruses were detected in 2017, which may be the reason for this, but this will be reported elsewhere.

Two thirds of the WNV-positive deaths were due to elective euthanasia, which was presumably due to either a grave prognosis or economic constraints affecting treatment. The extremely short median survival time until elective euthanasia was likely attributable to the rapid onset of severe clinical signs. Case fatality and neuroinvasive proportions in 2016–2017 were similar to those previously reported both in RSA [12] and internationally [13,14,55].

The most prevalent clinical signs displayed by cases during 2016–2017 were very similar to those described for equine WNV-positive patients in general, consisting mainly of various neurological signs with or without fever [5,6,12,27,55,56]. As with the previous ZARV study [12], neurological signs were present in most WNV-positive cases, and all neurological clinical signs were significantly associated with WNV infection ($p$ = 0.001). A consistently high percentage of neuroinvasion was seen among all age groups that tested WNV-positive. In both WNV cases and controls, subjects with only neurological signs had the highest odds of fatality, which is likely due to the severity of pathology in the CNS. It was interesting that the fatal WNV cases had higher proportions of hindleg and total paralysis than foreleg paralysis and tremors or muscle fasciculations, relating to the degree and location of spinal cord pathology. A significant, consistent decrease both in absolute WNV case numbers and in odds of WNV infection relative to other causes of neurological and febrile disease, was seen with increasing age ($p$ = 0.041), which was possibly due to an increased immunological resistance resulting from repeated low-grade exposure to WNV [5,12,13]. This is in contrast to human WNV cases in whom clinical signs are more apparent and severe in very young and very old patients, with neurological signs especially in the elderly [7,43,57]. The literature differs in opinion regarding whether the age of horses influences the case fatality rate or whether geriatric horses are more or less susceptible [13,58], but it is generally agreed that horses younger than 5 years are more susceptible to clinical disease. In rodents, increased viraemia correlated with earlier neuroinvasion and increased WNV burdens in the CNS [43]. Studies using South African lineage 2 strains showed that neuroinvasive LD50 levels for most strains were comparable to those of lineage 1 strains in the USA, although some strains were more neuroinvasive than others [59].

Fever, when present, was presumably due to a systemic inflammatory reaction to the WNV infection [43]. Approximately half of the WNV-positive cases in 2016–2017 had fever compared to 35% in 2008–2015 [12]. As in human patients [7], uncomplicated equine WNV cases usually recovered fully. The tendency for WNV cases with fever, with or without neurological signs, to be more likely to recover than those without fever was surprising. This can also be seen in another study that detailed the clinical presentation of lineage 2 WNV cases in horses in Austria in 2016 to 2018 [27]. Studies have shown that numerous attributes of the innate and adaptive immunity, particularly T-cell mediated immunity, are required to successfully counteract the viraemia and mitigate pathogenesis in the CNS [43]. The presence of fever may be an indication of the horse's ability to mount an effective general immune response to the initial viraemia. Therefore, the presence of certain clinical signs, such as fever, could potentially serve as a prognostic indicator for recovery.

Laminitic stance/sensitivity in feet was an interesting clinical sign noted in a small but statistically significant number of cases ($p = 0.023$). It is not generally described in the literature as a sign of equine WNV infection, although one study mentioned "reluctance to move" as a clinical sign [1] and another mentioned "hindlimb lameness/monoplegia" [27]. Some human WNV patients may experience severe pain in their limbs just before or during the onset of weakness [7], suggesting a neuropathic cause for the perceived pedal sensitivity in some of the equine patients. All five of these horses also displayed neurological signs, and two were eventually euthanised. Retained clinical signs such as gait or behavioural abnormality after recovery were seen in a small, marginally significant number of cases ($p = 0.053$), although there were much fewer than reported elsewhere [13,14]. It is possible that there would have been a higher proportion of retained clinical signs in South African horses if fewer economic constraints and more awareness of the disease had allowed a longer treatment period.

The various mixed breeds and the South African breeds, Boerperd and Nooitgedachter, were associated with a significantly lower odds of WNV infection compared to purebred and intermediate hybrid vigour groups ($p = 0.009$). These horses may show greater immunological resistance to endemic diseases due either to hybrid vigour or to some form of genetic adaptation, as seen in humans with certain gene mutations [60] or asymptomatic WNV infection in birds from endemic countries such as RSA [10]. Accurate data are not available, but a large proportion of the general South African equine population is likely comprised of non-purebred indigenous/local or mixed breeds; they were also well represented in the randomly selected controls, indicating that they were not overlooked due to being of less economic value.

Stabling at night, sex of horse, vaccination against African horse sickness, and being generally highly stressed in the four to six weeks before clinical symptoms were detected were not found to be significantly associated with WNV infection in the univariate analysis. It is important to note that a large proportion of the subjects (both cases and controls) had, according to the owners, experienced high levels of stress, particularly due to long-distance travelling, in the 4–6 weeks before sample submission. Research in rodents showed that increased stress levels promoted immunosuppression, increased WNV replication in vivo, and increased neuroinvasion causing encephalitis and death [20,61]. Considering that all horses in this study, including controls, were diseased, the effect of stress on WNV infection could not be evaluated; however, this suggests that certain stressors may play a role in equine disease development in general.

Being vaccinated against EIV may have been acting as a proxy for other unmeasured variable(s) associated with WNV infection risk, leading to its being retained in the final model as a confounder. EIV vaccination is compulsory for competitive sport horses, required for participation in events organised by the South African Equestrian Federation (SAEF) and other individual discipline associations such as the National Horse Racing Authority (NHRA). Competitive horses vaccinated against EIV would typically be more likely to travel than horses that are not vaccinated against EIV. Thus, it is possible that the stress of competing, travelling, or exposure of an immunologically naïve horse to WNV or to different regional strains of WNV, or another associated factor, may predispose a horse to contracting WNV disease.

Based on the follow-up of owners of affected horses, a much higher proportion vaccinated their herds after sample submission in 2017 to 2018 compared to 2016 to 2017. However, this number only reflects the study subjects from the CVZ database and is not necessarily an indication of increased vaccination proportion countrywide. Although vaccine sales data are not available, personal communication with one of the two pharmaceutical companies currently selling a WNV vaccine in RSA has confirmed that sales of the WNV vaccine increased during 2018 to 2019. This was noted particularly after the period of telephonic follow-up interviews with the owners and veterinarians, during which WNV infection and available vaccinations were explained. Awareness campaigns in recent years have also created public recognition of the potential for WNV to cause severe neurological disease and death in horses and humans in RSA and of the need for vaccination, which should result in decreased numbers of WNV-positive cases in horses in South Africa. However, due to economic constraints, or perhaps ignorance regarding immunology, one may expect many South African horse owners to neglect regular vaccination during years when few cases occur, as is the case with other non-government regulated vaccines. This may contribute to a decline in immunity, especially in younger, immunologically naïve horses, leading to increased WNV case numbers during outbreaks. There is also uncertainty amongst horse owners regarding the duration of natural immunity post-infection, creating differing opinions regarding whether horses should be vaccinated.

## 4. Materials and Methods

Ethical approval: The study protocol (V080-18), as well as previous testing of CVZ samples (H01216), was approved by the University of Pretoria Animal Ethics Committee and the Faculty of Veterinary Science Research Committee. Department of Agriculture, Land Reform and Rural Development (DALRRD) approval had been obtained for testing of CVZ samples under Section 20 approval. Annual reports of the cases detected by the CVZ were submitted to the DALRRD. Veterinarians and owners involved were informed of the purpose of the questionnaire and assented either verbally or electronically, as well as by written permission on the test requisition form, to the information being used for research purposes.

Study design: A case-control study was conducted using submissions to the ZARV programme during 2016 to 2017. Criteria for submission to the study were one or more of the following: fever (more than 38.5 °C) and/or acute neurological disease, and/or death. Neurological disease was characterised as horses displaying one or more of the following clinical signs: ataxia, blindness, facial paralysis, hyperreactivity or hyperaesthesia, incoordination, nystagmus, paresis, partial or complete paralysis, recumbency, seizures, tremors and muscle fasciculations, tongue paralysis and/or weakness, lip twitching, head tilting, and/or dysphagia.

Cases were defined as the horses in the ZARV database that tested positive for WNV infection on real-time reverse transcriptase polymerase chain reaction (rtRT-PCR) using the method published in [62], immunoglobulin M enzyme-linked immunosorbent assay (IgM ELISA) (IDEXX IgM kit, Hoofddorp, The Netherlands), or both. IgM-positive cases were confirmed by serum neutralisation assays in the BSL3 laboratory at the CVZ. All 54 available WNV-positive cases were used, of which 6 horses were diagnosed WNV-positive in 2016 and 48 were diagnosed WNV-positive in 2017. Most of the WNV-positive cases were diagnosed using IgM ELISA (47/54, 87%) of which two cases (4%) tested positive on both real-time reverse transcription polymerase chain reaction (rtRT-PCR) and ELISA. Only 17% (9/54) of the cases tested positive for WNV on rtRT-PCR. For each case, at least two WNV-negative control horses were randomly selected from the same population of ZARV sample submissions. These 120 control horses also complied with one or more of the three inclusion criteria. Any incomplete information on the sample requisition forms submitted for the subjects was obtained from the veterinarians who sent the samples or the owners of the horses. Interviews with owners also assessed general awareness of WNV, common clinical signs, potential pathological course, occurrence and distribution in RSA, and availability and use of vaccines. Additional information was obtained during telephonic interviews regarding the general vaccination status of the horses, potential stressors in the four to six weeks prior to displaying clinical signs (such as long-distance travelling, change of ownership and management system, recent illness or injury, weaning or recent African horse sickness vaccination), housing at night, and outcome of the disease.

Since some of the owners had multiple submissions to the dataset (for instance at a stud), responses regarding WNV vaccination were also grouped into herds based on location rather than individual horses. A total of 149 herds were interviewed (cases and controls), of which 47 herds contributed to the WNV-positive cases.

Horses were also grouped according to age; distribution of the WNV-positive cases was as follows: 56% ($n$ = 30) of cases were juveniles less than 5 years old, 39% were adult horses between the ages of 5 and 15 years old ($n$ = 21), and 5% were considered geriatric, being older than 15 years ($n$ = 3).

Data Analysis: Comparisons between years were done using all available data from the 583 CVZ sample submissions of horses during 2016 and 2017. Analysis of the case-control study was done using the 174 subjects on which complete data had been obtained. Variables for analysis were divided into 2 groups: the exposure or potential risk factors plausibly associated with the likelihood of a horse being WNV-positive were used to develop a multiple logistic regression model. Secondly, the recorded clinical signs and outcomes (partial/full recovery, death/euthanasia) were described and correlated with the presence or absence of WNV infection by means of univariate analysis.

Univariate analysis of risk factors was performed using cross-tabulation and two-tailed Fisher's exact test. For continuous variables, the assumption of linearity was assessed by plotting the Pearson and Deviance residuals against the value of the predictor in a simple logistic regression model and evaluating the linearity of the resulting pattern. The predictors were also categorised into quartiles, and the quartile midpoints were plotted against their estimated log odds to evaluate linearity. Elevation above sea-level (altitude) of the locations of WNV-positive cases was categorised into quartiles due to being non-linear. Some categorical variables were recoded to increase statistical power by combining categories with few observations.

Multivariable analysis was done using multiple logistic regression. For the initial logistic regression model, all univariate risk factor variables with the likelihood ratio test (LRT) $p < 0.2$ were included, and the least statistically significant variables were eliminated using a backward stepwise procedure. Finally, all independent variables were re-included one by one and retained if significant, or if inclusion resulted in substantial (>20%) changes in the coefficients for other variables in the model. Goodness of fit was evaluated using the Hosmer–Lemeshow goodness-of-fit statistic. Analysis was done using Stata 15 (StataCorp, College Station, TX, USA) and NCSS 2007 (NCSS, Kaysville, UT, USA). Significance was set at $p < 0.05$.

## 5. Conclusions

Increased equine WNV-positive case numbers in RSA in 2017 were largely attributed to environmental factors favouring the breeding habits of the vector. The largest proportion of cases during 2016–2017 was reported in the temperate to warm, eastern inland

RSA plateau, at intermediate elevation above sea level, during March–April. The WNV-associated case fatality rate and neuroinvasive disease proportions from 2016 to 2017 were consistent with those reported in previous local and international studies. Most of the cases displayed neurological signs, which were significantly associated with WNV infection, and approximately half of the cases had fever. Fever was marginally associated with recovery from WNV and may potentially be used as a prognostic indicator. Vaccination against WNV was significantly protective, and the risk of developing clinical WNV significantly decreased with increasing age, which was likely due to increased immunity from repeated long-term, low grade field exposure.

Therefore, it is advisable that owners with competitive horses or those younger than two to five years old, especially the highly purebred breeds (such as Thoroughbreds, Warmbloods, and Arabians) residing in the eastern temperate to warm parts of RSA with high summer rainfall, or travelling between provinces, should practice routine, complete vaccination against WNV. These vaccines should be given annually during spring, in order to decrease morbidity and mortality by timeously increasing immunological resistance against WNV.

**Author Contributions:** All three authors contributed to the conceptualisation, methodology, and resources for the study. Investigation and data curation were performed by F.-M.B. and M.V., while F.-M.B. and P.N.T. were responsible for the analysis and mapping software, formal analysis and visualisation. F.-M.B. prepared and wrote the original draft, and M.V. and P.N.T. supervised the project, validated the results, reviewed and edited the manuscript and contributed to funding acquisition. M.V. is the owner of the database and all specimens that had been tested for arboviruses under the Zoonotic Arbo and Respiratory Viruses Programme which had been approved under section 20 approval by the Department Agriculture Forestry and Fisheries and the University of Pretoria Animal Ethics Committee. All authors have read and agreed to the published version of the manuscript.

**Funding:** This research including WNV testing and laboratory investigations were funded through M.V.'s National Research Foundation incentive fund and University of Pretoria Development fund during 2016-2017 and were done at no extra expense to either the owner or veterinarian.

**Institutional Review Board Statement:** The study was conducted according to the ARRIVE guidelines and approved by the Animal Ethics Committee of the University of Pretoria (V080-18, 9 October 2018; H012-16, 25 July 2016). The study was also conducted according to the guidelines of the Declaration of Helsinki, and approved by the University of Pretoria, Faculty of Health ethics committee (protocol code 155/2019, 12 April 2019). Title: "One Health approach to detect zoonotic arboviruses through surveillance in animals and vectors, and development of molecular and serological assays to define their epidemiology in South Africa".

**Informed Consent Statement:** All animal specimens were submitted to the surveillance program under informed consent that the data will be used as part of arbovirus surveillance and research. The study did not involve human subjects thus human informed consent was not applicable.

**Data Availability Statement:** The data were collected as part of the research and surveillance program by the ZARV programme in the Centre for Viral Zoonoses, University of Pretoria. The data is not publicly available to protect the identity of the owners. The minimal dataset that supports the central findings of a published study may be obtained from the corresponding author, M.V. under agreement with the University of Pretoria.

**Acknowledgments:** The authors would like to express sincere appreciation to the owners, stud and stable managers, and veterinarians who provided information, and condolences to those who lost a beloved horse. We would also like to thank and acknowledge all staff in the Zoonotic Arbo and Respiratory Viruses Programme in the CVZ who participated in laboratory testing, as well as June Williams of the Department of Veterinary Pathology, University of Pretoria, who was personally involved in post-mortem examination and test requisition of many of the WNV-positive cases.

**Conflicts of Interest:** The authors declare no conflict of interest.

## References


1. Siger, L.; Bowen, R.; Karaca, K.; Murray, M.; Jagannatha, S.; Echols, B.; Nordgren, R.; Minke, J.M. Evaluation of the efficacy provided by a Recombinant Canarypox-Vectored Equine West Nile Virus vaccine against an experimental West Nile Virus intrathecal challenge in horses. *Vet. Ther.* **2006**, *7*, 249–256. [PubMed]
2. Venter, M.; Human, S.; Zaayman, D.; Gerdes, G.; Williams, J.; Steyl, J.; Leman, P.A.; Paweska, J.; Setzkorn, H.; Rous, G.; et al. Lineage 2 West Nile Virus as Cause of Fatal Neurologic Disease in Horses, South Africa. *Emerg. Infect. Dis.* **2009**, *15*, 877–884. [CrossRef] [PubMed]
3. Beasley, D.W.C.; Barrett, A.D.T.; Tesh, R.B. Resurgence of West Nile neurologic disease in the United States in 2012: What happened? What needs to be done? *Antiviral Res.* **2013**, *99*, 1–5. [CrossRef] [PubMed]
4. Petersen, L.R.; Roehrig, J.T. West Nile Virus: A Reemerging Global Pathogen. *Emerg. Infect. Dis.* **2001**, *7*, 611. [CrossRef] [PubMed]
5. Castillo-Olivares, J.; Wood, J. West Nile virus infection of horses. *Vet. Res.* **2004**, *35*, 467–483. [CrossRef]
6. OIE. Manual of Diagnostic Tests and Vaccines for Terrestrial Animals: West Nile Fever. Available online: http://www.oie.int/fileadmin/Home/eng/Health_standards/tahm/2.01.24_WEST_NILE.pdf (accessed on 3 August 2018).
7. Petersen, L.R.; Brault, A.C.; Nasci, R.S. West Nile virus: Review of the literature. *JAMA* **2013**, *310*, 308–315. [CrossRef]
8. Ciota, A.T. West Nile virus and its vectors. *Curr. Opin. Insect Sci.* **2017**, *22*, 28–36. [CrossRef]
9. Roehrig, J.T. West nile virus in the United States—A historical perspective. *Viruses* **2013**, *5*, 3088–3108. [CrossRef]
10. Jupp, P.G. The ecology of West Nile virus in South Africa and the occurrence of outbreaks in humans. *Ann. N. Y. Acad. Sci.* **2001**, *951*, 143–152. [CrossRef]
11. Burt, F.J.; Grobbelaar, A.A.; Leman, P.A.; Anthony, F.S.; Gibson, G.V.F.; Swanepoel, R. Phylogenetic relationships of southern African West Nile virus isolates. *Emerg. Infect. Dis.* **2002**, *8*, 820–826. [CrossRef]
12. Venter, M.; Pretorius, M.; Fuller, J.A.; Botha, E.; Rakgotho, M.; Stivaktas, V.; Weyer, C.; Romito, M.; Williams, J. West Nile Virus Lineage 2 in Horses and Other Animals with Neurologic Disease, South Africa, 2008–2015. *Emerg. Infect. Dis.* **2017**, *23*, 2060. [CrossRef] [PubMed]
13. AAEP. Core Vaccination Guidelines: West Nile Virus. Available online: https://aaep.org/guidelines/vaccination-guidelines/core-vaccination-guidelines/west-nile-virus (accessed on 27 March 2019).
14. Ward, M.P.; Schuermann, J.A.; Highfield, L.D.; Murray, K.O. Characteristics of an outbreak of West Nile virus encephalomyelitis in a previously uninfected population of horses. *Vet. Microbiol.* **2006**, *118*, 255–259. [CrossRef] [PubMed]
15. Lanciotti, R.S.; Ebel, G.D.; Deubel, V.; Kerst, A.J.; Murri, S.; Meyer, R.; Bowen, M.; McKinney, N.; Morrill, W.E.; Crabtree, M.B.; et al. Complete genome sequences and phylogenetic analysis of West Nile virus strains isolated from the United States, Europe, and the Middle East. *Virology* **2002**, *298*, 96–105. [CrossRef] [PubMed]
16. Charrel, R.N.; Brault, A.C.; Gallian, P.; Lemasson, J.J.; Murgue, B.; Murri, S.; Pastorino, B.; Zeller, H.; de Chesse, R.; de Micco, P.; et al. Evolutionary relationship between Old World West Nile virus strains. Evidence for viral gene flow between Africa, the Middle East, and Europe. *Virology* **2003**, *315*, 381–388. [CrossRef]
17. Bakonyi, T.; Hubalek, Z.; Rudolf, I.; Nowotny, N. Novel flavivirus or new lineage of West Nile virus, central Europe. *Emerg. Infect. Dis.* **2005**, *11*, 225–231. [CrossRef] [PubMed]
18. Vazquez, A.; Sanchez-Seco, M.P.; Ruiz, S.; Molero, F.; Hernandez, L.; Moreno, J.; Magallanes, A.; Tejedor, C.G.; Tenorio, A. Putative new lineage of west nile virus, Spain. *Emerg. Infect. Dis.* **2010**, *16*, 549–552. [CrossRef]
19. Lanciotti, R.S.; Roehrig, J.T.; Deubel, V.; Smith, J.; Parker, M.; Steele, K.; Crise, B.; Volpe, K.E.; Crabtree, M.B.; Scherret, J.H.; et al. Origin of the West Nile Virus Responsible for an Outbreak of Encephalitis in the Northeastern United States. *Science* **1999**, *286*, 2333. [CrossRef]
20. John, H.R.; Scott, R.D.; Zdenek, H. Migratory Birds and Spread of West Nile Virus in the Western Hemisphere. *Emerg. Infect. Dis.* **2000**, *6*, 319. [CrossRef]
21. Barrett, A.D.T. West Nile in Europe: An increasing public health problem. *J. Travel Med.* **2018**, *25*, tay096. [CrossRef]
22. CDC. West Nile Virus Final Cumulative Maps & Data for 1999–2019 in North America. Available online: https://www.cdc.gov/westnile/statsmaps/cumMapsData.html (accessed on 14 December 2020).
23. Sambri, V.; Capobianchi, M.; Charrel, R.; Fyodorova, M.; Gaibani, P.; Gould, E.; Niedrig, M.; Papa, A.; Pierro, A.; Rossini, G.; et al. West Nile virus in Europe: Emergence, epidemiology, diagnosis, treatment, and prevention. *Clin. Microbiol. Infect.* **2013**, *19*, 699–704. [CrossRef]
24. Di Sabatino, D.; Bruno, R.; Sauro, F.; Danzetta, M.L.; Cito, F.; Iannetti, S.; Narcisi, V.; De Massis, F.; Calistri, P. Epidemiology of West Nile disease in Europe and in the Mediterranean Basin from 2009 to 2013. *Biomed. Res. Int.* **2014**, *2014*, 907852. [CrossRef] [PubMed]
25. ECDC. Surveillance and Disease Data for West Nile Fever. Available online: https://www.ecdc.europa.eu/en/west-nile-fever/surveillance-and-disease-data/disease-data-ecdc (accessed on 14 December 2020).
26. Camp, J.V.; Nowotny, N. The knowns and unknowns of West Nile virus in Europe: What did we learn from the 2018 outbreak? *Expert Rev. Anti-Infect. Ther.* **2020**, *18*, 145–154. [CrossRef] [PubMed]
27. de Heus, P.; Kolodziejek, J.; Camp, J.V.; Dimmel, K.; Bagó, Z.; Hubálek, Z.; van den Hoven, R.; Cavalleri, J.-M.V.; Nowotny, N. Emergence of West Nile virus lineage 2 in Europe: Characteristics of the first seven cases of West Nile neuroinvasive disease in horses in Austria. *Transbound. Emerg. Dis.* **2020**, *67*, 1189–1197. [CrossRef] [PubMed]

28. Steyn, J.; Botha, E.; Stivaktas, V.I.; Buss, P.; Beechler, B.R.; Myburgh, J.G.; Steyl, J.; Williams, J.; Venter, M. West Nile Virus in Wildlife and Nonequine Domestic Animals, South Africa, 2010–2018. *Emerg. Infect. Dis.* **2019**, *25*, 2290–2294. [CrossRef] [PubMed]
29. Venter, M.; Human, S.; Van Niekerk, S.; Williams, J.; van Eeden, C.; Freeman, F. Fatal neurologic disease and abortion in mare infected with lineage 1 West Nile virus, South Africa. *Emerg. Infect. Dis.* **2011**, *17*, 1534–1536. [CrossRef] [PubMed]
30. Venter, M.; Swanepoel, R. West Nile virus lineage 2 as a cause of zoonotic neurological disease in humans and horses in Southern Africa. *Vector Borne Zoonotic Dis.* **2010**, *10*, 659–664. [CrossRef]
31. Williams, J.H.; van Niekerk, S.; Human, S.; van Wilpe, E.; Venter, M. Pathology of fatal lineage 1 and 2 West Nile virus infections in horses in South Africa. *J. S. Afr. Vet. Assoc.* **2014**, *85*, 1105. [CrossRef]
32. McIntosh, B.M.; Jupp, P.G.; Dos Santos, I.; Meenehan, G.M. Epidemics of West Nile and Sindbis viruses in South Africa with Culex (Culex) univittatus Theobald as vector. *S. Afr. J. Sci.* **1976**, *72*, 295–300.
33. Sule, W.F.; Oluwayelu, D.O.; Hernández-Triana, L.M.; Fooks, A.R.; Venter, M.; Johnson, N. Epidemiology and ecology of West Nile virus in sub-Saharan Africa. *Parasites Vectors* **2018**, *11*, 414. [CrossRef]
34. Zaayman, D.; Venter, M. West Nile Virus Neurologic Disease in Humans, South Africa, September 2008–May 2009. *Emerg. Infect. Dis.* **2012**, *18*, 2051. [CrossRef]
35. Botha, E.M.; Markotter, W.; Wolfaardt, M.; Paweska, J.T.; Swanepoel, R.; Palacios, G.; Nel, L.H.; Venter, M. Genetic determinants of virulence in pathogenic lineage 2 West Nile virus strains. *Emerg. Infect. Dis.* **2008**, *14*, 222–230. [CrossRef] [PubMed]
36. Van Eeden, C.; Swanepoel, R.; Venter, M. Antibodies against West Nile and Shuni viruses in veterinarians, South Africa. *Emerg. Infect. Dis.* **2014**, *20*, 1409–1411. [CrossRef] [PubMed]
37. OIE. WAHIS Interface Zoonotic Diseases in Humans: West Nile Virus in South Africa. Available online: https://www.oie.int/wahis_2/public/wahid.php/Countryinformation/Zoonoses#ZAF (accessed on 14 December 2020).
38. Jupp, P.G. Mosquitoes as vectors of human disease in South Africa. *S. Afr. Fam. Pract.* **2005**, *47*, 68–72. [CrossRef]
39. Cornel, A.J.; Jupp, P.G.; Blackburn, N.K. Environmental temperature on the vector competence of Culex univittatus (Diptera: Culicidae) for West Nile virus. *J. Med. Entomol.* **1993**, *30*, 449–456. [CrossRef] [PubMed]
40. DeFelice, N.B.; Schneider, Z.D.; Little, E.; Barker, C.; Caillouet, K.A.; Campbell, S.R.; Damian, D.; Irwin, P.; Jones, H.M.P.; Townsend, J.; et al. Use of temperature to improve West Nile virus forecasts. *PLoS Comp. Biol.* **2018**, *14*, e1006047. [CrossRef] [PubMed]
41. Guthrie, A.J.; Howell, P.G.; Gardner, I.A.; Swanepoel, R.E.; Nurton, J.P. West Nile virus infection of Thoroughbred horses in South Africa (2000–2001). *Equine Vet. J.* **2003**, *35*, 601–605. [CrossRef] [PubMed]
42. Van Niekerk, S.; Human, S.; Williams, J.; van Wilpe, E.; Pretorius, M.; Swanepoel, R.; Venter, M. Sindbis and Middelburg Old World Alphaviruses Associated with Neurologic Disease in Horses, South Africa. *Emerg. Infect. Dis.* **2015**, *21*, 2225. [CrossRef]
43. Samuel, M.A.; Diamond, M.S. Pathogenesis of West Nile Virus infection: A balance between virulence, innate and adaptive immunity, and viral evasion. *J. Virol.* **2006**, *80*, 9349–9360. [CrossRef]
44. Cantile, C.; Di Guardo, G.; Eleni, C.; Arispici, M. Clinical and neuropathological features of West Nile virus equine encephalomyelitis in Italy. *Equine Vet. J.* **2000**, *32*, 31–35. [CrossRef]
45. Long, M.T.; Ostlund, E.N.; Porter, M.B.; Crom, R.L. Equine West Nile encephalitis: Epidemiological and clinical review for practitioners. In Proceedings of the 48th Annual Convention of the American Association of Equine Practitioners, Orlando, FL, USA, 4–8 December 2002; pp. 1–6.
46. Seino, K.K.; Long, M.T.; Gibbs, E.P.J.; Bowen, R.A.; Beachboard, S.E.; Humphrey, P.P.; Dixon, M.A.; Bourgeois, M.A. Comparative Efficacies of Three Commercially Available Vaccines against West Nile Virus (WNV) in a Short-Duration Challenge Trial Involving an Equine WNV Encephalitis Model. *Clin. Vaccine Immunol.* **2007**, *14*, 1465. [CrossRef]
47. Siger, L.; Bowen, R.A.; Karaca, K.; Murray, M.J.; Gordy, P.W.; Loosmore, S.M.; Audonnet, J.C.; Nordgren, R.M.; Minke, J.M. Assessment of the efficacy of a single dose of a recombinant vaccine against West Nile virus in response to natural challenge with West Nile virus-infected mosquitoes in horses. *Am. J. Vet. Res.* **2004**, *65*, 1459–1462. [CrossRef]
48. Venter, M.; van Vuren, P.J.; Mentoor, J.; Paweska, J.; Williams, J. Inactivated West Nile Virus (WNV) vaccine, Duvaxyn WNV, protects against a highly neuroinvasive lineage 2 WNV strain in mice. *Vaccine* **2013**, *31*, 3856–3862. [CrossRef] [PubMed]
49. DeFelice, N.B.; Little, E.; Campbell, S.R.; Shaman, J. Ensemble forecast of human West Nile virus cases and mosquito infection rates. *Nat. Commun.* **2017**, *8*, 14592. [CrossRef] [PubMed]
50. SAWS. Historical Rain Maps. Available online: http://www.weathersa.co.za/Home/HistoricalRain (accessed on 25 March 2019).
51. Hayes, E.B.; Komar, N.; Nasci, R.S.; Montgomery, S.P.; O'Leary, D.R.; Campbell, G.L. Epidemiology and transmission dynamics of West Nile virus disease. *Emerg. Infect. Dis.* **2005**, *11*, 1167–1173. [CrossRef] [PubMed]
52. Jupp, P.G.; Blackburn, N.K.; Thompson, D.L.; Meenehan, G.M. Sindbis and West Nile virus infections in the Witwatersrand-Pretoria region. *S. Afr. Med. J.* **1986**, *70*, 218–220.
53. SAWS. *Weathersmart News February 2018*; South African Weather Service: Centurion, Gauteng, South Africa, 2018; p. 29.
54. Hartley, C. *Personal Communication: Thoroughbred Stud Farm Survey 2017*; The Thoroughbred Breeders' Association of South Africa: Germiston, South Africa, 2019.
55. WHO. West Nile Virus Fact Sheet. Available online: http://www.who.int/news-room/fact-sheets/detail/west-nile-virus (accessed on 28 November 2017).
56. Weese, J.S. AAEP Infectious Disease Guidelines: West Nile Virus. Available online: https://aaep.org/sites/default/files/Documents/WestNileVirus_Final.pdf (accessed on 9 April 2019).

57. Taylor, R.M.; Work, T.H.; Hurlbut, H.S.; Rizk, F. A Study of the Ecology of West Nile Virus in Egypt. *Am. J. Trop. Med. Hyg.* **1956**, *5*, 579–620. [CrossRef]
58. Epp, T.; Waldner, C.; West, K.; Townsend, H. Factors associated with West Nile virus disease fatalities in horses. *Can. Vet. J.* **2007**, *48*, 1137–1145.
59. Venter, M.; Myers, T.G.; Wilson, M.A.; Kindt, T.J.; Paweska, J.T.; Burt, F.J.; Leman, P.A.; Swanepoel, R. Gene expression in mice infected with West Nile virus strains of different neurovirulence. *Virology* **2005**, *342*, 119–140. [CrossRef]
60. Petersen, L.R.; Marfin, A.A. West Nile virus: A primer for the clinician. *Ann. Intern. Med.* **2002**, *137*, 173–179. [CrossRef]
61. Ben-Nathan, D. Stress and Virulence: West Nile Virus Encephalitis. *Isr. J. Vet. Med.* **2013**, *68*, 6.
62. Zaayman, D.; Human, S.; Venter, M. A highly sensitive method for the detection and genotyping of West Nile virus by real-time PCR. *J. Virol. Methods* **2009**, *157*, 155–160. [CrossRef] [PubMed]

*pathogens*

MDPI

*Article*

# First Detection of the West Nile Virus Koutango Lineage in Sandflies in Niger

**Gamou Fall [1,*], Diawo Diallo [2], Hadiza Soumaila [3,4], El Hadji Ndiaye [2], Adamou Lagare [5], Bacary Djilocalisse Sadio [1], Marie Henriette Dior Ndione [1], Michael Wiley [6,7], Moussa Dia [1], Mamadou Diop [8], Arame Ba [1], Fati Sidikou [5], Bienvenu Baruani Ngoy [9], Oumar Faye [1], Jean Testa [5], Cheikh Loucoubar [8], Amadou Alpha Sall [1], Mawlouth Diallo [2] and Ousmane Faye [1]**

1 Pole of Virology, WHO Collaborating Center For Arbovirus and Haemorrhagic Fever Virus, Institut Pasteur, Dakar BP 220, Senegal; Bacary.SADIO@pasteur.sn (B.D.S.); Marie.NDIONE@pasteur.sn (M.H.D.N.); Moussa.DIA@pasteur.sn (M.D.); Arame.ba@pasteur.sn (A.B.); Oumar.FAYE@pasteur.sn (O.F.); Amadou.SALL@pasteur.sn (A.A.S.); Ousmane.FAYE@pasteur.sn (O.F.)
2 Pole of Zoology, Medical Entomology Unit, Institut Pasteur, Dakar BP 220, Senegal; Diawo.DIALLO@pasteur.sn (D.D.); elhadji.ndiaye@pasteur.sn (E.H.N.); mawlouth.diallo@pasteur.sn (M.D.)
3 Programme National de Lutte contre le Paludisme, Ministère de la Santé Publique du Niger, Niamey BP 623, Niger; hadiza_soumaila@pmivectorlink.com
4 PMI Vector Link Project, Niamey BP 11051, Niger
5 Centre de Recherche Médicale et Sanitaire, Niamey BP 10887, Niger; lagare@cermes.org (A.L.); fati@cermes.org (F.S.); jean.testa@unice.fr (J.T.)
6 United States Army Medical Research Institute of Infectious Diseases, Fort Detrick, MD 21702-5011, USA; mike.wiley@unmc.edu
7 Department of Environmental, Agricultural, and Occupational Health, University of Nebraska, Omaha, NE 68198-4355, USA
8 Biostatistic, Biomathematics and Modelling Group, Institut Pasteur, Dakar BP 220, Senegal; mamadou.diop@pasteur.sn (M.D.); Cheikh.Loucoubar@pasteur.sn (C.L.)
9 WHO Country Office, Niamey B.P. 10739, Niger; baruaningoyb@who.int
* Correspondence: gamou.fall@pasteur.sn; Tel.: +221-338399200; Fax: +221-338399210

**Citation:** Fall, G.; Diallo, D.; Soumaila, H.; Ndiaye, E.H.; Lagare, A.; Sadio, B.D.; Ndione, M.H.D.; Wiley, M.; Dia, M.; Diop, M.; et al. First Detection of the West Nile Virus Koutango Lineage in Sandflies in Niger. *Pathogens* **2021**, *10*, 257. https://doi.org/10.3390/pathogens10030257

Academic Editor: Andrew Taylor-Robinson

Received: 15 January 2021
Accepted: 19 February 2021
Published: 24 February 2021



**Abstract:** West Nile virus (WNV), belonging to the *Flaviviridae* family, causes a mosquito-borne disease and shows great genetic diversity, with at least eight different lineages. The Koutango lineage of WNV (WN-KOUTV), mostly associated with ticks and rodents in the wild, is exclusively present in Africa and shows evidence of infection in humans and high virulence in mice. In 2016, in a context of Rift Valley fever (RVF) outbreak in Niger, mosquitoes, biting midges and sandflies were collected for arbovirus isolation using cell culture, immunofluorescence and RT-PCR assays. Whole genome sequencing and in vivo replication studies using mice were later conducted on positive samples. The WN-KOUTV strain was detected in a sandfly pool. The sequence analyses and replication studies confirmed that this strain belonged to the WN-KOUTV lineage and caused 100% mortality of mice. Further studies should be done to assess what genetic traits of WN-KOUTV influence this very high virulence in mice. In addition, given the risk of WN-KOUTV to infect humans, the possibility of multiple vectors as well as birds as reservoirs of WNV, to spread the virus beyond Africa, and the increasing threats of flavivirus infections in the world, it is important to understand the potential of WN-KOUTV to emerge.

**Keywords:** West Nile virus; Koutango lineage; high virulence; sandflies; Niger

## 1. Introduction

West Nile virus (WNV) is flavivirus maintained in nature through an enzootic transmission cycle between *Culex* spp. mosquitoes including *Cx. pipiens*, *Cx. quinquefasciatus*, *Cx. neavei* and birds [1–3]. WN fever outbreaks occur essentially in humans and horses, considered as dead-end hosts [4]. Clinical symptoms range from asymptomatic or flu-like illness to severe neurological and meningoencephalitis syndromes [3]. WNV is one of the most widespread flaviviruses worldwide, has caused massive human and animal infections, and some fatal cases, particularly in America and Europe [5–8]. WNV has a great genetic diversity with at least eight different lineages, and of them four (lineages 1, 2, Koutango and putative new lineage 8) are present in Africa [9]. The WN lineage 1 is distributed worldwide and has been responsible for all major WN outbreaks [6,8,10,11]. The WN lineage 2 was exclusively present in Africa until 2004, when it emerged in Europe and replaced lineage 1 [12]. Migratory birds that overwintered in Africa were the most likely source of introduction of WN lineage 2 into Europe. The putative new lineage was isolated from *Cx. perfuscus* in the south-east of Senegal in 1992 and was never found associated with animals or humans [9]. The Koutango lineage of West Nile virus (WN-KOUTV) was first isolated in 1968 from the wild rodent *Tatera kempi* in Senegal [13] and in 1974 from gerbils in Somalia [14]. WN-KOUTV was initially classified as a distinct virus and later, based on phylogenetic studies, considered a WNV lineage [15,16]. WN-KOUTV is exclusively detected in Africa, and unlike other WNV lineages, it was once isolated from mosquitoes and mainly from ticks and rodents, particularly in Senegal [17]. In humans, serological evidence in Gabon [18] and a report of an accident where a Senegalese laboratory worker was symptomatically infected with WN-KOUTV [19] have been shown. Different symptoms such as two-day fever accompanied by achiness and retrobulbar headache, to erythematous eruption on the flanks, were detected [13,19]. Patients with acute febrile illness ruled out for malaria and Lassa fever in Sierra Leone were found to present neutralizing antibodies to WN-KOUTV [20]. This unpublished study shows that natural human infections with WN-KOUTV are occurring in Africa and suggests that this virus is likely the etiological agent of at least some of the fevers with unknown origin.

In animal models, intra-cerebral inoculation of the virus to new-borne mice causes death on days three to four post-infection [13], and in adult mice, WN-KOUTV showed higher virulence compared to all other WNV lineages [17,21,22]. Currently, the transmission cycle remains unclear, and the roles of ticks and mosquitoes are not yet known. Indeed, vector competence studies showed that *Cx. quinquefasciatus* and *Cx. neavei*, proven vectors for other WNV lineages, were not competent for WN-KOUTV lineage [9]. Another vector competence study showed that *Aedes aegypti* was found to carry the virus and disseminated the infection after a blood meal with high viral dose [23], suggesting that this mosquito species could probably transmit WN-KOUTV lineage to humans. It has also been shown that *Ae. aegypti* was able to vertically transmit the virus [24]. Vector competence of ticks, mostly associated with the virus in the wild, has never been tested.

In the context of Rift Valley fever (RVF) outbreak investigations in Niger, in 2016 [25], mosquitoes and sandflies were collected for arbovirus detection. The laboratory analyses revealed the presence of the WN-KOUTV lineage in a pool of sandflies. Here, we describe this first detection of WN-KOUTV in sandflies, the virus isolation, viral genome sequencing and analyses, and in vivo characterization in mice.

## 2. Results

### *2.1. Arthropod Species*

A total of 10,977 hematophagous arthropods (158 mosquitoes, 10,816 sandflies and 3 biting midges) were collected and grouped into 181 pools (Table 1). Of these, sandflies were the only arthropod group collected in Intoussane, and the most abundant in Tchintabaraden ($n$ = 359; 83.3%) and Tasnala ($n$ = 10,428; 99.2%). *Anopheles gambiae* (58.2%) and *Cx perexiguus* (24.7%) were the most abundant among six mosquito species collected during our collection period. Sandflies were collected only by Centers for Disease Control and Prevention (CDC) light traps near herds at Intoussane, and ground pools at Tasnala, while they were found in all biotopes investigated at Tchintabaraden.

**Table 1.** List of arthropods collected in the field from three districts of Niger, 20–24 October 2016.

| Species | Tchintabaraden | Intoussane | Tasnala | Total |
|---|---|---|---|---|
| *Anopheles gambiae* | 40 | | 52 | 92 |
| *Anopheles rufipes* | 1 | | 3 | 4 |
| *Cx. antennatus* | 1 | | | 1 |
| *Cx. ethiopicus* | 1 | | 3 | 4 |
| *Cx. quinquefasciatus* | 14 | | 4 | 18 |
| *Cx. perexiguus* | 14 | | 25 | 39 |
| Total mosquitoes | 71 | | 87 | 158 |
| Biting midges | 1 | | 2 | 3 |
| Sandflies | 359 | 29 | 10,428 | 10,816 |
| Total arthropods | 431 | 29 | 10,517 | 10,977 |

### 2.2. Virus Detection

One pool of 100 sandflies collected by CDC light trap near a ground pool at Tasnala was positive for flavivirus by immunofluorescence assay IFA and West Nile virus by real time RT-PCR. The genotyping using WNV primers and probes specific to the different lineages, as well as the genome sequencing, showed the presence of WN-KOUTV in the sample.

The minimum infection rate (MIR) was 0.01 per 1000 in the ground pool where the positive pool was collected.

### 2.3. Sequencing and Evolutionary Analyses

A genome sequence of 10,948 bp was obtained. The BLAST search showed that the sequence corresponds to WN-KOUTV and the Genbank accession number is MN057643. Phylogenetic analyses showed that the sequence of the strain isolated from sandflies belongs to the same cluster as other WN-KOUTV strains, ArD96655 (accession number KY703855.1) and Dak Ar D 5443 (accession number EU082200.2). This analysis confirmed, therefore, that the virus strain from sandflies in Niger belongs to the WN-KOUTV lineage (Figure 1).

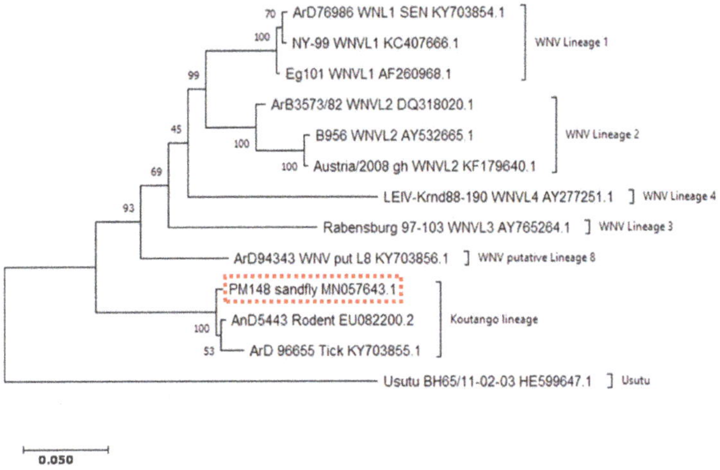


**Figure 1.** Phylogenetic analyses using MEGA software and the maximum likelihood method. Phylogenetic analyses were conducted with sequences of the sandfly strain, other Koutango strains isolated in Senegal from ticks and rodents, and other WNV lineage strains. The accession numbers and names of the strains are mentioned. The accession number of the sandfly strain genomic sequence, PM148 from Niger, is MN057643. WNVL1: lineage 1, WNVL2: lineage 2, WNVL3: lineage 3, WNVL4: lineage 4 and WNV Put L8: putative lineage 8.

Amino acid sequence analyses of the sandfly strain and other WNV strains showed a mean genetic distance around 0.01 within the Koutango lineage (Table 2). The sandfly strain was more closed to the rodent WN-KOUTV strain, with a genetic distance of 0.005. As with other WN-KOUTV strains, the sandfly strain also showed high genetic distances with other WNV lineages, ranging from 11 to 18% (Table 2).

Amino acid sequence alignment of the new sandfly strain and other WNV strains was performed to check for mutations that have been shown to impact WNV virulence (Figure 2). Mutations already described for WN-KOUTV strains [26–28] have been detected in the pre-membrane (S to M at position 72), envelope (Y to F at position 155 of the glycosylation site), and NS5 (F to S at position 653) proteins of the new sandfly WN-KOUTV strain. In addition, the rodent WN-KOUTV strain showed a specific mutation (S to P) at position 156 of the envelope protein glycosylation site. Mutations with unknown consequence (SVA to ASS), specific to all WN-KOUTV strains analyzed here, were also detected at positions 363 to 365 of the envelope protein. The new sandfly WN-KOUTV strain shared a mutation (A to T) with WNVL1 and L2 at position 366 of the envelope protein and showed specific/unique mutations compared to other WN-KOUTV strains at position 38 (M to I) of the NS2A and 177 (V to M) of the NS3 proteins (Figure 2).

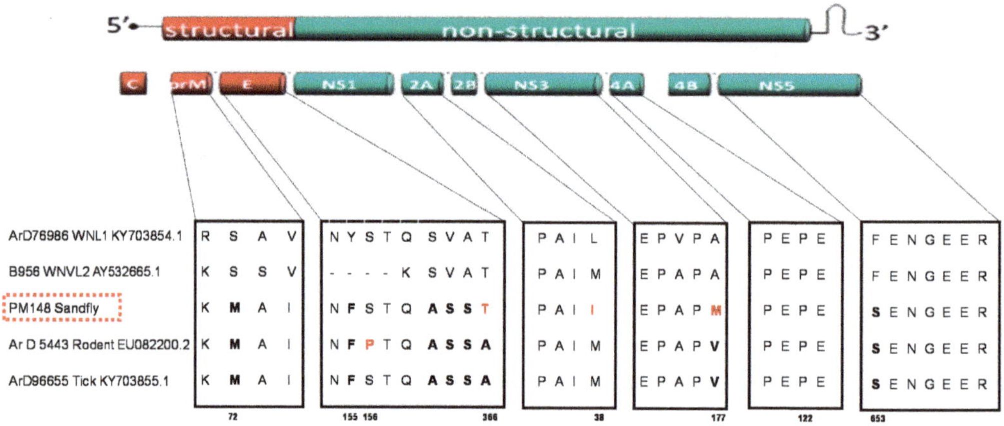


**Figure 2.** Genetic diversity of WNV lineages 1, 2 and Koutango. Alignment was conducted with sequences of the sandfly strain, other Koutango strains isolated in Senegal from ticks and rodents, and other WNV lineages 1 and 2 strains. The genomic structure of West Nile virus is shown, and the different genes are labeled. Alignments of motifs with unknown consequence and known virulence motifs are shown. Mutations specific to all Koutango strains are in bold and mutations specific to one Koutango strain are in red.

### 2.4. *In Vivo Characterization*

Intra-cerebral inoculation of the new KOUTV strain isolated from sandflies to newborne mice showed 100% mortality at day two post-infection, while ArD96655 showed 100% mortality at day four post-infection. In adult mice, the Koutango sandfly strain showed 100% mortality of mice at days 7 and 10 post-infection with 100 and 1000 pfu, respectively, while ArD96655 showed 100% mortality at day six post-infection with both doses (Figure 3).

**Table 2.** Genetic distance analysis conducted using MEGA software with the Poisson correction model.

| Strains | 1 | 2 | 3 | 4 | 5 | 6 | 7 | 8 | 9 | 10 | 11 | 12 |
|---|---|---|---|---|---|---|---|---|---|---|---|---|
| 1.PM148_sandfly_MN057643.1_Koutango_lineage | | | | | | | | | | | | |
| 2.AnD5443_Rodent_EU082200.2_Koutango_lineage | 0.00584 | | | | | | | | | | | |
| 3.ArD_96655_Tick_KY703855.1_Koutango_lineage | 0.02121 | 0.00848 | | | | | | | | | | |
| 4.ArD76986_SEN_KY703854.1_WNV_Lineage_1 | 0.11371 | 0.11469 | 0.11764 | | | | | | | | | |
| 5.Eg101_AF260968.1_WNV_Lineage_1 | 0.13990 | 0.11208 | 0.14274 | 0.00584 | | | | | | | | |
| 6.NY-99_KC407666.1_WNV_Lineage_1 | 0.14174 | 0.11404 | 0.14398 | 0.00467 | 0.00769 | | | | | | | |
| 7.ArB3573/82_DQ318020.1_WNV_Lineage_2 | 0.13645 | 0.11371 | 0.14039 | 0.06095 | 0.07205 | 0.07557 | | | | | | |
| 8.Austria/2008_gh_KF179640.1_WNV_Lineage_2 | 0.13084 | 0.11310 | 0.12682 | 0.06004 | 0.08429 | 0.08696 | 0.03494 | | | | | |
| 9.B956_AY532665.1_WNV_Lineage_2 | 0.13294 | 0.11353 | 0.12892 | 0.06196 | 0.08603 | 0.08930 | 0.03663 | 0.00634 | | | | |
| 10.ArD94343_KY703856.1_WNV_putative_Lineage_8 | 0.12553 | 0.12752 | 0.12917 | 0.09528 | 0.09432 | 0.09368 | 0.09176 | 0.09147 | 0.09156 | | | |
| 11.LEIV-Krnd88-190_AY277251.1_WNV_Lineage_4 | 0.18058 | 0.15722 | 0.18384 | 0.12070 | 0.13969 | 0.14094 | 0.13333 | 0.13487 | 0.13640 | 0.14568 | | |
| 12.Rabensburg_97-103_AY765264.1_WNV_Lineage_3 | 0.16703 | 0.14386 | 0.17689 | 0.10107 | 0.12331 | 0.12359 | 0.11968 | 0.12823 | 0.13222 | 0.12224 | 0.16220 | |

In grey, genetic distances within WN-KOUTV lineage. Red rectangle, genetic distances between sandfly strain and with other WNV lineages.

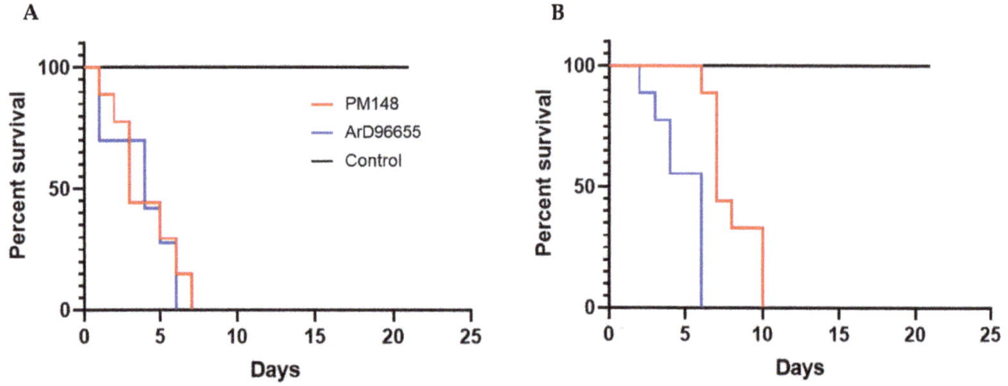


**Figure 3.** Survival curves of 5- to 6-week-old mice following intraperitoneal infection with (**A**) 100, and (**B**) 1000 pfu. Eight mice were tested for each dose. A group of mice with an injection of PBS was used as control. Mice were monitored daily for 21 days.

In both experiments, PBS-inoculated negative control groups showed no signs of disease and stayed alive throughout the experiments.

## 3. Discussion

Our study showed, for the first time to the best of our knowledge, the isolation of WN-KOUTV from sandflies but also the detection of this particular West Nile virus lineage in Niger. Indeed, phylogenetic analyses showed that the virus strain from sandflies exhibited similar genotypic patterns to other WN-KOUTV strains already described [9,17]. WN-KOUTV was previously isolated once from mosquitoes and several times from ticks in Senegal, then this isolation in sandflies extended the spectrum of the potential WN-KOUTV vectors and highlighted once again the particular feature of this WNV lineage. The vector competence of ticks and sandflies, naturally associated with the WN-KOUTV lineage, is not proven, but in laboratory conditions it has been shown that *Ae. aegypti* can transmit the WN-KOUTV lineage, but only with a high viral dose in an artificial blood meal. This suggests that only high viremia will naturally render a vertebrate host infectious for the *Ae. aegypti* mosquito [24]. However, little is known about WN-KOUTV infections and viremia titers in vertebrate hosts. In addition, apart from rodents, the existence of other vertebrate hosts in the wild is not known. Obviously, because WN-KOUTV is a WNV lineage, birds might also play important roles in its transmission as well as propagation beyond the African continent, as proposed for lineages 1 and 2 [2,6]. All these considerations emphasize the need to better characterize this particular WNV lineage in Africa and its epidemic potential. Therefore, more studies are needed to help understand the potential of the common mosquito species *Ae. aegypti* to transmit naturally WN-KOUTV between different vertebrate hosts, and the role of birds in the transmission cycle. Vector competence studies of ticks and sandflies species are also necessary to better understand the transmission dynamics of WN-KOUTV.

Sequence analyses conducted in this study showed high genetic distances between WN-KOUTV and other WNV lineages, which confirmed that Koutango is the most distant WNV lineage [17]. The sequence alignment showed variations specific to KOUTV lineage and also between KOUTV strains. The mutations found in the envelope, pre-membrane and NS5, in all WN-KOUTV strains, could therefore explain the higher virulence of this lineage compared to other WNV lineages. In addition, although high virulence was observed for both WN-KOUTV strains in mice, differences in the survival times of newborn and adult mice were noted. Further studies with more WN-KOUTV strains are therefore needed, to better characterize the genetic variations inside the WN-KOUTV lineage and their impact on the infection in vertebrate hosts. These studies will also help to understand if, like WNV lineages 1 and 2, strains with high and low virulence exist in the WN-KOUTV lineage.

No WN-KOUTV strain was detected in the different *Culex* spp. mosquitoes collected in this study. This could be partly explained by the low number of *Culex* spp. specimens collected. However, a previous vector competence study targeting two *Culex species,* considered as the most probable WNV vectors in domestic and enzootic contexts in Senegal, also showed that they were not competent for WN-KOUTV [9]. More vector competence studies targeting different vector species and WN-KOUTV strains are needed to better characterize the role of *Culex* mosquitoes in the transmission of WN-KOUTV.

In our study, the sandfly species positive for WN-KOUTV is not known because the sandflies collected during this investigation were not identified at species level. However, seven species including *Phlebotomus roubaudi, Phlebotomus clydei* and *Phlebotomus orientalis* were previously collected in Niger [29–31] and could be targeted for vector competence studies. The very high abundance of sandflies observed around dry pools in our study is concordant with previous investigations in the Sahelian area of Senegal [32] and Mauritania [33]. The peak abundance of these sandflies was shown to occur in the Sahelian area at the beginning of the dry season, around two months after the rain pools dry up [32]. Because they are only abundant during the dry season, sandflies would probably play a role in WN-KOUTV transmission in this period, while mosquitoes and/or ticks would be the main arthropods involved during the rainy season. Many other viruses were previously detected in phlebotomine sandflies in Africa, including Chandipura virus, Saboya virus, Tete virus, and two unknown viruses in Senegal [32,34], Yellow fever virus in Uganda [35], Perinet virus in Madagascar [36], sandfly fever Sicilian and sandfly fever Naples viruses in Egypt [37], and Punique virus in Tunisia [38]. Interestingly, like WN-KOUTV, Saboya virus, another flavivirus was also detected in both sandflies and rodents in Senegal [32,39]. This emphasizes the need to better characterize and evaluate the potential roles of sandflies in arbovirus transmission cycles, because many studies are only focused on mosquitoes and ticks.

## 4. Materials and Methods

### 4.1. Collection and Processing of Arthropods

Field investigations were conducted between 20–24 October 2016, in the villages of Tchintabaraden, Intoussane and Tasnala located in the district of Tchintabaraden (15°53′53″ N; 5°48′11″ E), Tahoua Region, Niger. These villages were selected based on the presence of confirmed human RVF cases and high abortion rates in ungulates. Hematophagous arthropods (mosquitoes, sandflies and biting midges) were collected in and around households of suspected and confirmed human RVF cases, herds, and the edges of ground pools using a backpack aspirator [40], CDC light traps [41], and indoor residual spaying [42].

Arthropods collected were frozen, morphologically identified to the species level for mosquitoes using morphological keys [43,44], and family level for other arthropods, pooled by family, species, sex, and date. All arthropod pools were conserved at the laboratory in Niamey, and later aliquots were transported to Institut Pasteur de Dakar for virus testing and isolation. The minimum infection rate (MIR) was calculated to estimate the viral infection rate in arthropod populations by assuming that at least one individual of the pooled sample could be infected. The formula of MIR is as follows; MIR = number of positive pools/numbers of tested arthropods × 1000.

### 4.2. Virus Isolation

The arthropod pools were homogenized in 3 mL of L-15 medium (Gibco BRL, Grand Island, NY, USA) supplemented with 20% of fetal bovine serum and clarified by centrifugation at 1500× *g*, at 4 °C for 10 min. The supernatants were then filtered using a 1 mL syringe (Artsana, Como, Italy) and sterilized with 0.20 µm filters (Sartorius, Göttingen, Germany).

Viral isolation was conducted from the supernatants using C6-36 (Ae. albopictus) cells, and the presence of virus was detected by immunofluorescence assay (IFA) using in-house immune ascite pools specific to different flaviviruses, bunyaviruses, orbiviruses, and alphaviruses, as previously described [45].

### 4.3. RT-PCR and Titration

RNA extraction was conducted from the supernatant of the IFA-positive sample using the QiaAmp Viral RNA Extraction Kit (Qiagen, Heiden, Germany) according to the manufacturer's instructions. The RNA samples were then screened by RT-PCR (reverse transcription-polymerase chain reaction) for flaviviruses (dengue, yellow fever, Zika and West Nile virus). The primers and probes already described elsewhere were used [46–49].

The supernatant of the IFA-positive sample (confirmed by RT-PCR) was titrated as previously described, using PS cells (Porcine Stable kidney cells, ATCC number, Manassas, VA, USA) [50].

### 4.4. Viral Sequencing and Phylogenetic Analyses

Host ribosomal RNAs were depleted prior to sequencing from extracted RNA, using specific probes from partners at the United States Army Medical Research Institute of Infectious Diseases (USAMRIID). The sequence-independent, single-primer amplification (SISPA) method was used for cDNA synthesis from depleted RNAs, and libraries were prepared using the Nextera XT library prep kit (Illumina, San Diego, CA, USA) with dual index strategy, according to the manufacturer's instructions. Libraries were normalized and pooled with PhiX DNA as loading control, and the sequencing was performed using Miseq, Illumina, for 2 × 151 cycles. Sequencing runs were monitored in real time using the Illumina Sequencing Viewer Analyzer for cluster density, percentage of clusters passing filter, phasing/pre-phasing ratios, % base, error rates, % reads with quality score ≥30, and other parameters. The bioinformatics analyses were performed via an in-house script that implements a pipeline for pathogen discovery. De novo assembly was performed using Geneious prime v. 2019.1.3 to obtain the complete sequence. A BLAST search was then conducted to identify the assembled sequence.

Alignment and evolutionary analyses (genetic distance and phylogenetic analyses) were conducted with amino acid sequences using MEGA-X 10.2.2 software (MEGA, University of Pennsylvania, USA). Genetic distance analysis was conducted using the Poisson correction model and the phylogenetic analysis was inferred by using the maximum likelihood method and JTT (Jones, Taylor, and Thornton) matrix-based model with a bootstrap test of 1000 replicates [51,52]. The tree with the highest log likelihood is shown in Figure 1.

### 4.5. In Vivo Characterization in Mice

The IFA-positive sample was tested in newborn and adult mice in comparison with the KOUTV strain ArD96655 (accession number KY703855.1) isolated from ticks.

Ten new-borne Swiss mice (1–2 days old) were inoculated with 1000 pfu of IFA positive sample, ArD96655 and PBS alone, by an intra-cerebral route and were monitored daily for 21 days.

Five- to six-week-old Swiss mice were also challenged by intraperitoneal route with 100 pfu and 1000 pfu, to analyze the virulence of the IFA-positive sample. The WN-KOUTV strain ArD96655 and PBS were also tested as positive and negative controls, respectively. Eight mice were tested for each dose. Survival curves were generated using GraphPad Prism 9.0.0 (121) software (GraphPad, San Diego, CA, USA).

## 5. Conclusions

In this study, we have shown the circulation of WN-KOUTV in Niger and its detection in sandflies for the first time. These results extended the number of countries in Africa where this virus is reported, but also the spectrum of potential vectors. The very high virulence in mice [17,21,22], the possibility of multiple vectors, the risk of KOUTV to infect humans, and the increasing threats of flavivirus infections in the world, should contribute to better consideration of WNV-KOUTV as an important emerging pathogen. In this regard, vector competence studies, vertebrate hosts including birds, and viral genetic diversity characterizations are ongoing and will provide new insights on WN-KOUTV transmission, virulence and possibility to diffuse beyond Africa. Seroprevalence studies would also be important in Niger, particularly in Tahoua Region where the virus was isolated, and in Senegal where WN-KOUTV was isolated several times, to assess the potential circulation of KOUTV in humans.

**Author Contributions:** Conceptualization, G.F., D.D., M.D. (Mawlouth Diallo), O.F. (Ousmane Faye), and A.A.S.; methodology, E.H.N., D.D., H.S., G.F., B.D.S., and M.W.; formal analysis, M.W., M.D. (Mamadou Diop), M.H.D.N., and G.F.; validation, D.D., G.F., A.L., M.D. (Moussa Dia), M.H.D.N., C.L., and M.W.; investigation, E.H.N., D.D., H.S., A.L., G.F., M.H.D.N., M.D. (Moussa Dia), A.B., and B.D.S.; resources, A.L., F.S., J.T., B.B.N.; M.W., O.F. (Ousmane Faye), M.D. (Mawlouth Diallo), and A.A.S., writing—original draft preparation, G.F., writing—review and editing, H.S., A.L., D.D., G.F., M.D. (Mawlouth Diallo), O.F. (Ousmane Faye), O.F. (Oumar Faye), M.W., B.B.N., J.T., C.L., and F.S.; supervision, F.S., G.F., D.D., B.D.S., and A.L. All authors have read and agreed to the published version of the manuscript.

**Funding:** The field investigations in Niger were funded by WHO AFRO and the laboratory work was funded by Institut Pasteur de Dakar. This research received no external funding.

**Institutional Review Board Statement:** There is no National Ethics Committee for animals in Senegal. These studies were performed in the context of surveillance at the WHO collaborating center for arboviruses and hemorrhagic fever viruses, and all experiments on animals were conducted by respecting the World Organization for Animal Health regulations (https://www.oie.int/fileadmin/Home/fr/Health_standards/tahc/current/chapitre_aw_research_education.pdf) (accessed on 10 January 2021).

**Informed Consent Statement:** Not applicable.

**Data Availability Statement:** All data are included in the manuscript. The virus sequence generated during the current study is available at Genbank (accession number: MN057643).

**Acknowledgments:** The authors thank the authorities of the Ministry of Health of Niger and WHO Country office of Niger and Afro region for facilitating the Rift Valley fever outbreak investigation.

**Conflicts of Interest:** The authors declare no conflict of interest.

## References

1. Hubálek, Z.; Halouzka, J. West Nile Fever—A Reemerging Mosquito-Borne Viral Disease in Europe. *Emerg. Infect. Dis.* **1999**, *5*, 643–650. [CrossRef]
2. Rappole, J.H.; Hubálek, Z. Migratory Birds and West Nile Virus. *J. Appl. Microbiol.* **2003**, *94*, 47S–58S. [CrossRef]
3. Hayes, E.B.; Komar, N.; Nasci, R.S.; Montgomery, S.P.; O'Leary, D.R.; Campbell, G.L. Epidemiology and Transmission Dynamics of West Nile Virus Disease. *Emerg. Infect. Dis.* **2005**, *11*, 1167–1173. [CrossRef]
4. Kulasekera, V.L.; Kramer, L.; Nasci, R.S.; Mostashari, F.; Cherry, B.; Trock, S.C.; Glaser, C.; Miller, J.R. West Nile Virus Infection in Mosquitoes, Birds, Horses, and Humans, Staten Island, New York, 2000. *Emerg. Infect. Dis.* **2001**, *7*, 722–725. [CrossRef]
5. DeGroote, J.P.; Sugumaran, R.; Ecker, M. Landscape, Demographic and Climatic Associations with Human West Nile Virus Occurrence Regionally in 2012 in the United States of America. *Geospat. Health* **2014**, *9*, 153–168. [CrossRef]
6. Hernández-Triana, L.M.; Jeffries, C.L.; Mansfield, K.L.; Carnell, G.; Fooks, A.R.; Johnson, N. Emergence of West Nile Virus Lineage 2 in Europe: A Review on the Introduction and Spread of a Mosquito-Borne Disease. *Front. Public Health* **2014**, *2*, 271. [CrossRef] [PubMed]
7. Vilibic-Cavlek, T.; Kaic, B.; Barbic, L.; Pem-Novosel, I.; Slavic-Vrzic, V.; Lesnikar, V.; Kurecic-Filipovic, S.; Babic-Erceg, A.; Listes, E.; Stevanovic, V.; et al. First Evidence of Simultaneous Occurrence of West Nile Virus and Usutu Virus Neuroinvasive Disease in Humans in Croatia during the 2013 Outbreak. *Infection* **2014**, *42*, 689–695. [CrossRef] [PubMed]
8. Cox, S.L.; Campbell, G.D.; Nemeth, N.M. Outbreaks of West Nile Virus in Captive Waterfowl in Ontario, Canada. *Avian Pathol. J. WVPA* **2015**, *44*, 135–141. [CrossRef] [PubMed]
9. Fall, G.; Diallo, M.; Loucoubar, C.; Faye, O.; Sall, A.A. Vector Competence of Culex Neavei and Culex Quinquefasciatus (Diptera: Culicidae) from Senegal for Lineages 1, 2, Koutango and a Putative New Lineage of West Nile Virus. *Am. J. Trop. Med. Hyg.* **2014**, *90*, 747–754. [CrossRef]

10. Murray, K.O.; Mertens, E.; Despres, P. West Nile Virus and Its Emergence in the United States of America. *Vet. Res.* **2010**, *41*, 67. [CrossRef]
11. Anukumar, B.; Sapkal, G.N.; Tandale, B.V.; Balasubramanian, R.; Gangale, D. West Nile Encephalitis Outbreak in Kerala, India, 2011. *J. Clin. Virol.* **2014**, *61*, 152–155. [CrossRef] [PubMed]
12. Bakonyi, T.; Ivanics, E.; Erdélyi, K.; Ursu, K.; Ferenczi, E.; Weissenböck, H.; Nowotny, N. Lineage 1 and 2 Strains of Encephalitic West Nile Virus, Central Europe. *Emerg. Infect. Dis.* **2006**, *12*, 618–623. [CrossRef] [PubMed]
13. Coz, J.; Le Gonidec, G.; Cornet, M.; Valade, M.; Lemoine, M.; Gueye, A. Transmission Experimentale d'un Arbovirus Du Groupe B, Le Virus Koutango Par *Aedes aegypti* L. *Cah. ORSTOM Ser. Ent. Med. Parasitol.* **1975**, *13*, 57–62.
14. Butenko, A.M.; Semashko, I.V.; Skvortsova, T.M.; Gromashevskiĭ, V.L.; Kondrashina, N.G. Detection of the Koutango virus (Flavivirus, Togaviridae) in Somalia. *Med. Parazitol. (Mosk.)* **1986**, 65–68. Available online: https://pubmed.ncbi.nlm.nih.gov/3018465/ (accessed on 22 February 2021).
15. Charrel, R.N.; Brault, A.C.; Gallian, P.; Lemasson, J.-J.; Murgue, B.; Murri, S.; Pastorino, B.; Zeller, H.; de Chesse, R.; de Micco, P.; et al. Evolutionary Relationship between Old World West Nile Virus Strains. Evidence for Viral Gene Flow between Africa, the Middle East, and Europe. *Virology* **2003**, *315*, 381–388. [CrossRef]
16. Hall, R.A.; Scherret, J.H.; Mackenzie, J.S. Kunjin Virus: An Australian Variant of West Nile? *Ann. N. Y. Acad. Sci.* **2001**, *951*, 153–160. [CrossRef] [PubMed]
17. Fall, G.; Di Paola, N.; Faye, M.; Dia, M.; de Melo Freire, C.C.; Loucoubar, C.; de Andrade Zanotto, P.M.; Faye, O. Biological and Phylogenetic Characteristics of West African Lineages of West Nile Virus. *PLoS Negl. Trop. Dis.* **2017**, *11*, e0006078. [CrossRef]
18. Jan, C.; Languillat, G.; Renaudet, J.; Robin, Y. A serological survey of arboviruses in Gabon. *Bull. Soc. Pathol. Exot. Fil.* **1978**, *71*, 140–146.
19. Shope, R.E. Epidemiology of Other Arthropod-Borne Flaviviruses Infecting Humans. *Adv. Virus Res.* **2003**, *61*, 373–391. [CrossRef]
20. de Araujo Lobo, J. Koutango: Under Reported Arboviral Disease in West Africa. Ph.D. Thesis, Louisiana State University, Baton Rouge, LA, USA, 2012.
21. Pérez-Ramírez, E.; Llorente, F.; Del Amo, J.; Fall, G.; Sall, A.A.; Lubisi, A.; Lecollinet, S.; Vázquez, A.; Jiménez-Clavero, M.Á. Pathogenicity Evaluation of Twelve West Nile Virus Strains Belonging to Four Lineages from Five Continents in a Mouse Model: Discrimination between Three Pathogenicity Categories. *J. Gen. Virol.* **2017**, *98*, 662–670. [CrossRef]
22. Prow, N.A.; Setoh, Y.X.; Biron, R.M.; Sester, D.P.; Kim, K.S.; Hobson-Peters, J.; Hall, R.A.; Bielefeldt-Ohmann, H. The West Nile Virus-like Flavivirus Koutango Is Highly Virulent in Mice Due to Delayed Viral Clearance and the Induction of a Poor Neutralizing Antibody Response. *J. Virol.* **2014**, *88*, 9947–9962. [CrossRef]
23. de Araújo Lobo, J.M.; Christofferson, R.C.; Mores, C.N. Investigations of Koutango Virus Infectivity and Dissemination Dynamics in Aedes Aegypti Mosquitoes. *Environ. Health Insights* **2014**, *8*, 9–13. [CrossRef]
24. Coz, J.; Valade, M.; Cornet, M.; Robin, Y. Transovarian transmission of a Flavivirus, the Koutango virus, in *Aedes aegypti* L. *C. R. Hebd. Seances Acad. Sci. Ser. Sci. Nat.* **1976**, *283*, 109–110.
25. Lagare, A.; Fall, G.; Ibrahim, A.; Ousmane, S.; Sadio, B.; Abdoulaye, M.; Alhassane, A.; Mahaman, A.E.; Issaka, B.; Sidikou, F.; et al. First Occurrence of Rift Valley Fever Outbreak in Niger, 2016. *Vet. Med. Sci.* **2019**, *5*, 70–78. [CrossRef]
26. Beasley, D.W.C.; Whiteman, M.C.; Zhang, S.; Huang, C.Y.-H.; Schneider, B.S.; Smith, D.R.; Gromowski, G.D.; Higgs, S.; Kinney, R.M.; Barrett, A.D.T. Envelope Protein Glycosylation Status Influences Mouse Neuroinvasion Phenotype of Genetic Lineage 1 West Nile Virus Strains. *J. Virol.* **2005**, *79*, 8339–8347. [CrossRef] [PubMed]
27. Setoh, Y.X.; Prow, N.A.; Hobson-Peters, J.; Lobigs, M.; Young, P.R.; Khromykh, A.A.; Hall, R.A. Identification of Residues in West Nile Virus Pre-Membrane Protein That Influence Viral Particle Secretion and Virulence. *J. Gen. Virol.* **2012**, *93*, 1965–1975. [CrossRef] [PubMed]
28. Van Slyke, G.A.; Ciota, A.T.; Willsey, G.G.; Jaeger, J.; Shi, P.-Y.; Kramer, L.D. Point Mutations in the West Nile Virus (Flaviviridae; Flavivirus) RNA-Dependent RNA Polymerase Alter Viral Fitness in a Host-Dependent Manner in Vitro and in Vivo. *Virology* **2012**, *427*, 18–24. [CrossRef]
29. Parrot, L.; Hornet, J.; Cadenat, J. Note on the Phlebotomines. XLVIII. Phlebotomines of French West Africa. 1. Senegal, Soudan, Niger. *Arch. Inst. Pasteur D'Algerie* **1945**, *23*, 232–244.
30. Le Pont, F.; Robert, V.; Vattier-Bernard, G.; Rispail, P.; Jarry, D. Notes on the phlebotomus of Aïr (Niger). *Bull. Soc. Pathol. Exot. 1990* **1993**, *86*, 286–289.
31. Abonnenc, E.; Dyemkouma, A.; Hamon, J. On the presence of phlebotomus (phlebotomus) orientalis parrot, 1936, in the republic of niger. *Bull. Soc. Pathol. Exot. Fil.* **1964**, *57*, 158–164.
32. Fontenille, D.; Traore-Lamizana, M.; Trouillet, J.; Leclerc, A.; Mondo, M.; Ba, Y.; Digoutte, J.P.; Zeller, H.G. First Isolations of Arboviruses from Phlebotomine Sand Flies in West Africa. *Am. J. Trop. Med. Hyg.* **1994**, *50*, 570–574. [CrossRef] [PubMed]
33. Nabeth, P.; Kane, Y.; Abdalahi, M.O.; Diallo, M.; Ndiaye, K.; Ba, K.; Schneegans, F.; Sall, A.A.; Mathiot, C. Rift Valley Fever Outbreak, Mauritania, 1998: Seroepidemiologic, Virologic, Entomologic, and Zoologic Investigations. *Emerg. Infect. Dis.* **2001**, *7*, 1052–1054. [CrossRef] [PubMed]
34. Ba, Y.; Trouillet, J.; Thonnon, J.; Fontenille, D. Phlebotomine Sandflies Fauna in the Kedougou Area of Senegal, Importance in Arbovirus Transmission. *Bull. Soc. Pathol. Exot.* **1999**, *92*, 131–135. [PubMed]

35. Smithburn, K.C.; Haddow, A.J.; Lumsden, W.H.R. An Outbreak of Sylvan Yellow Fever in Uganda with Aëdes (Stegomyia) Africanus Theobald as Principal Vector and Insect Host of the Virus. *Ann. Trop. Med. Parasitol.* **1949**, *43*, 74–89. [CrossRef] [PubMed]
36. Clerc, Y.; Rodhain, F.; Digoutte, J.P.; Tesh, R.; Heme, G.; Coulanges, P. The Perinet virus, rhabdoviridae, of the vesiculovirus type isolated in Madagascar from Culicidae. *Arch. Inst. Pasteur Madag.* **1982**, *49*, 119–129.
37. Schmidt, J.R.; Schmidt, M.L.; Said, M.I. Phlebotomus Fever in Egypt. Isolation of Phlebotomus Fever Viruses from Phlebotomus Papatasi. *Am. J. Trop. Med. Hyg.* **1971**, *20*, 483–490. [CrossRef] [PubMed]
38. Zhioua, E.; Moureau, G.; Chelbi, I.; Ninove, L.; Bichaud, L.; Derbali, M.; Champs, M.; Cherni, S.; Salez, N.; Cook, S.; et al. Punique Virus, a Novel Phlebovirus, Related to Sandfly Fever Naples Virus, Isolated from Sandflies Collected in Tunisia. *J. Gen. Virol.* **2010**, *91*, 1275–1283. [CrossRef]
39. Saluzzo, J.F.; Adam, F.; Heme, G.; Digoutte, J.P. Isolation of viruses from rodents in Senegal (1983–1985). Description of a new poxvirus. *Bull. Soc. Pathol. Exot. Fil.* **1986**, *79*, 323–333.
40. Clark, G.G.; Seda, H.; Gubler, D.J. Use of the "CDC Backpack Aspirator" for Surveillance of Aedes Aegypti in San Juan, Puerto Rico. *J. Am. Mosq. Control. Assoc.* **1994**, *10*, 119–124. [PubMed]
41. Sudia, W.D.; Chamberlain, R.W. Battery-Operated Light Trap, an Improved Model. *J. Am. Mosq. Control Assoc.* **1988**, *4*, 536–538.
42. Service, M. *Mosquito Ecology: Field Sampling Methods*, Chapman Hall ed.; Springer: London, UK, 1993.
43. Edwards, E. *Mosquitoes of the Ethiopian Region: III Culicine Adults and Pupae*; British Museum Natural History: London, UK, 1941.
44. Diagne, N.; Fontenille, D.; Konate, L.; Faye, O.; Lamizana, M.T.; Legros, F.; Molez, J.F.; Trape, J.F. Anopheles of Senegal. An annotated and illustrated list. *Bull. Soc. Pathol. Exot.* **1994**, *87*, 267–277.
45. Digoutte, J.P.; Calvo-Wilson, M.A.; Mondo, M.; Traore-Lamizana, M.; Adam, F. Continuous Cell Lines and Immune Ascitic Fluid Pools in Arbovirus Detection. *Res. Virol.* **1992**, *143*, 417–422. [CrossRef]
46. Wagner, D.; de With, K.; Huzly, D.; Hufert, F.; Weidmann, M.; Breisinger, S.; Eppinger, S.; Kern, W.V.; Bauer, T.M. Nosocomial Acquisition of Dengue. *Emerg. Infect. Dis.* **2004**, *10*, 1872–1873. [CrossRef] [PubMed]
47. Weidmann, M.; Faye, O.; Faye, O.; Kranaster, R.; Marx, A.; Nunes, M.R.T.; Vasconcelos, P.F.C.; Hufert, F.T.; Sall, A.A. Improved LNA Probe-Based Assay for the Detection of African and South American Yellow Fever Virus Strains. *J. Clin. Virol.* **2010**, *48*, 187–192. [CrossRef] [PubMed]
48. Faye, O.; Faye, O.; Diallo, D.; Diallo, M.; Weidmann, M.; Sall, A.A. Quantitative Real-Time PCR Detection of Zika Virus and Evaluation with Field-Caught Mosquitoes. *Virol. J.* **2013**, *10*, 311. [CrossRef] [PubMed]
49. Fall, G.; Faye, M.; Weidmann, M.; Kaiser, M.; Dupressoir, A.; Ndiaye, E.H.; Ba, Y.; Diallo, M.; Faye, O.; Sall, A.A. Real-Time RT-PCR Assays for Detection and Genotyping of West Nile Virus Lineages Circulating in Africa. *Vector Borne Zoonotic Dis.* **2016**, *16*, 781–789. [CrossRef] [PubMed]
50. De Madrid, A.T.; Porterfield, J.S. A Simple Micro-Culture Method for the Study of Group B Arboviruses. *Bull. World Health Organ.* **1969**, *40*, 113–121.
51. Jones, D.T.; Taylor, W.R.; Thornton, J.M. The Rapid Generation of Mutation Data Matrices from Protein Sequences. *Comput. Appl. Biosci. CABIOS* **1992**, *8*, 275–282. [CrossRef]
52. Kumar, S.; Stecher, G.; Li, M.; Knyaz, C.; Tamura, K. MEGA X: Molecular Evolutionary Genetics Analysis across Computing Platforms. *Mol. Biol. Evol.* **2018**, *35*, 1547–1549. [CrossRef]

*pathogens*

MDPI

*Article*

# Contrasted Epidemiological Patterns of West Nile Virus Lineages 1 and 2 Infections in France from 2015 to 2019

**Cécile Beck [1,*], Isabelle Leparc Goffart [2,3], Florian Franke [4], Gaelle Gonzalez [1], Marine Dumarest [1], Steeve Lowenski [1], Yannick Blanchard [5], Pierrick Lucas [5], Xavier de Lamballerie [3], Gilda Grard [2,3], Guillaume André Durand [2,3], Stéphan Zientara [1], Jackie Tapprest [6], Grégory L'Ambert [7], Benoit Durand [8], Stéphanie Desvaux [9,†] and Sylvie Lecollinet [1,†]**

[1] UMR 1161 Virology, ANSES, INRAE, ENVA, ANSES Animal Health Laboratory, EURL for Equine Diseases, 94704 Maisons-Alfort, France; gaelle.gonzalez@anses.fr (G.G.); marine.dumarest@anses.fr (M.D.); steeve.lowenski@anses.fr (S.L.); stephan.zientara@anses.fr (S.Z.); sylvie.lecollinet@anses.fr (S.L.)

[2] National Reference Laboratory for Arboviruses, Institut de Recherche Biomédicale des Armées, 13010 Marseille, France; isabelle.leparc-goffart@inserm.fr (I.L.G.); Gilda.Grard@inserm.fr (G.G.); guillaume.durand@inserm.fr (G.A.D.)

[3] Unité des Virus Emergents (UVE: Aix Marseille Univ, IRD 190, INSERM 1207, IHU Méditerranée Infection), 13385 Marseille, France; xavier.de-lamballerie@univ-amu.fr

[4] Regional Office Paca-Corse, Santé Publique France, 13331 Marseille, France; Florian.franke@santepubliquefrance.fr

[5] ANSES Ploufragan-Plouzane-Niort, Viral Genetics and Biosecurity Unit, 22440 Ploufragan, France; Yannick.blanchard@anses.fr (Y.B.); pierrick.lucas@anses.fr (P.L.)

[6] ANSES Animal Health Laboratory, PhEED Unit, Dozulé site, F-14430 Goustranville, France; Jackie.tapprest@anses.fr

[7] EID Méditerranée, 34184 Montpellier, France; glambert@eid-med.org

[8] Epidemiology Unit, Paris-Est University, ANSES Animal Health Laboratory, 94704 Maisons-Alfort, France; benoit.durand@anses.fr

[9] Office Français de la Biodiversité, Unité Sanitaire de la Faune, 01330 Birieux, France; stephanie.desvaux@ofb.gouv.fr

* Correspondence: cecile.beck@anses.fr; Tel.: +33-143-96-73-34

† These authors contributed equally to this work.

Received: 5 October 2020; Accepted: 28 October 2020; Published: 30 October 2020

**Abstract:** Since 2015, annual West Nile virus (WNV) outbreaks of varying intensities have been reported in France. Recent intensification of enzootic WNV circulation was observed in the South of France with most horse cases detected in 2015 ($n = 49$), 2018 ($n = 13$), and 2019 ($n = 13$). A WNV lineage 1 strain was isolated from a horse suffering from West Nile neuro-invasive disease (WNND) during the 2015 episode in the Camargue area. A breaking point in WNV epidemiology was achieved in 2018, when WNV lineage 2 emerged in Southeastern areas. This virus most probably originated from WNV spread from Northern Italy and caused WNND in humans and the death of diurnal raptors. WNV lineage 2 emergence was associated with the most important human WNV epidemics identified so far in France (n = 26, including seven WNND cases and two infections in blood and organ donors). Two other major findings were the detection of WNV in areas with no or limited history of WNV circulation (Alpes-Maritimes in 2018, Corsica in 2018–2019, and Var in 2019) and distinct spatial distribution of human and horse WNV cases. These new data reinforce the necessity to enhance French WNV surveillance to better anticipate future WNV epidemics and epizootics and to improve the safety of blood and organ donations.

**Keywords:** arbovirus; emerging infectious diseases; zoonotic; West Nile; lineages 1 and 2; France

## 1. Introduction

West Nile virus (WNV) is an arthropod-borne flavivirus transmitted by the bites of infected mosquitoes mostly belonging to the *Culex* genus [1]. WNV is maintained in an enzootic bird-mosquito cycle but can be transmitted through mosquito bites to dead-end hosts, such as humans or equids and occasionally cause neuro-invasive disease that can be lethal in these hosts [2]. In Europe, WNV outbreaks occur during the summer and fall seasons (July–October) when *Culex* mosquitoes are abundant.

According to phylogenetic analysis, eight different WNV lineages have been described [3]. WNV lineages 1 and 2 are the most widespread and caused most of the major epidemics encountered so far [4]. WNV lineage 1 was first reported in Europe in the 1960s when seropositive animals (horses and cattle) or viral isolates (mosquitoes and humans) were identified in France, Portugal, and Cyprus [5,6]. After more than 30 years without experiencing WNV outbreaks, North African, Western, and Eastern European countries reported again the emergence of WNV lineage 1 strains belonging to the Western-Mediterranean clade (in North Africa and Western Europe such as in Morocco in 1996, Italy in 1998, and in France in 2000) [7–9] and the Eastern-European clade (Eastern Europe during the 1996 Romanian outbreak and Russia in 1999) [10] affecting mainly equids and or/humans respectively. In France, after the 2000 WNV outbreaks in the Camargue area, sporadic cases of West Nile fever (WNF) occurred in departments bordering the Mediterranean coast in the 2000s (Var in 2003, Bouches-Du-Rhône, Gard and Hérault in the Camargue area in 2004, and Eastern Pyrenees in 2006) [11,12]. WNF has been mainly reported in France in horses, while human cases have been less frequently observed (seven cases in Var in 2003) [13].

Most European WNV outbreaks before 2010 had been caused by WNV lineage 1 strains. Unexpectedly, recent increase in WNV transmission and outbreaks, noticeable in Europe since 2010, has been associated with the introduction and spread of WNV lineage 2 strains [14–16]. A first WNV lineage 2 strain was initially detected in Hungary in 2004 [17] and subsequently spread to the eastern part of Austria in 2008 [15,18], to the Balkan peninsula, including Greece in 2010 [19], Serbia, Croatia, and Bulgaria in 2012 [20,21], further East to Italy in 2011 [22], and more recently it reached Spain in 2017 [23] and Germany in 2018 [24,25]. Another WNV lineage 2 strain, first detected in 2004 in Rostov Oblast in Southern Russia [26], has also been occasionally reported in Europe, in Romania [27] in 2010, in Italy in 2014 [28], and in Greece in 2018 [29].

After WNV reemergence in the Camargue area in 2000, a multidisciplinary WNV monitoring system has been implemented in France since 2001 including clinical surveillance in wild birds, horses, and humans, and records of mosquito abundance and diversity during the transmission season. Such clinical surveillance is implemented in departments from the Mediterranean area during the WNV at-risk period from 1 June until the end of November [11].

Here we report recent French WNV outbreaks (2015–2019) in humans, equids, and the wild avifauna in the Mediterranean area and describe the emergence of WNV lineage 2 in France in 2018 and the changing patterns of infection in humans and horses following this emergence.

## 2. Results

### 2.1. Comparison of WNV Seasonal Patterns in France between 2015 and 2019

#### 2.1.1. WNV Outbreaks in France in 2015–2019

In France during the last five years, the most important equine WNV outbreaks were reported in 2015. Of the 49 cases reported in horses in 2015, 41 presented neuroinvasive forms, three febrile forms, and five asymptomatics were identified thanks to serosurveys implemented in identified WNV transmission foci [11]. There was no equine case noticed in 2016, while one asymptomatic horse was detected in 2017 in the vicinity of WNV human cases. A total of 13 and nine neuroinvasive cases were reported in 2018 and 2019 respectively, while four additional febrile forms were reported in 2019 (Figure 1). Natural death or euthanasia occurred in 14.6% (6/41) and 15.4% (2/13) of the horses with

West Nile neuroinvasive disease (WNND) in 2015 and 2018 respectively whereas no equine death was reported in 2019. These percentages are lower than usually described in the literature [30].

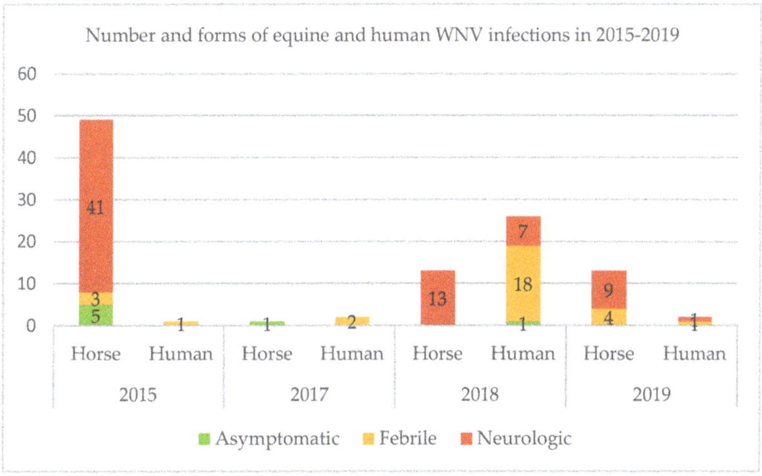


**Figure 1.** A graph that shows the total number and clinical forms of human and equine laboratory-confirmed cases per year in France from 2015 to 2019. The highest number of human and equine cases for the 2015–2019 period was reported in 2018 and 2015 respectively.

For the same period, for humans, the highest number of autochthonous cases was reported in 2018 with 26 laboratory-diagnosed human cases, including seven WNND, 18 febrile and one asymptomatic form. Interestingly, one blood donor, symptomatic a few days after the donation and one organ asymptomatic donor were tested positive for WNV the same year [31]. The total number of cases in 2018 represented a 27-fold increase compared with the 2015–2017 transmission seasons during which one, none, and two febrile cases were reported in 2015, 2016, and 2017 respectively. In 2019, one febrile and one WNND were reported (Figure 1). WNND have generally occurred more frequently in horses than in humans in France so far.

For the first time in France since the implementation of an integrated WNV surveillance system, WNV infections were also reported through WNV surveillance in the avifauna in 2018. In total, four raptors (two northern goshawks (*Accipiter gentilis*), one common buzzard (*Buteo buteo*), and one long-eared owl (*Asio otus*) were diagnosed WNV positive in September and October 2018 in Corsica (owl) and Alpes-Maritimes (diurnal raptors) by the SAGIR network (a French network dedicated to wildlife disease surveillance). All these wild birds were found alive and suffered from serious nervous disorders.

### 2.1.2. Shifts in Temporal and Spatial Distribution of WNV Cases

In 2015, the onset of the first equine case was reported on week 33 (starting the 11 August) with a peak on week 38 (14 September to 20 September) and the last case was notified on week 44 (Figure 2a).

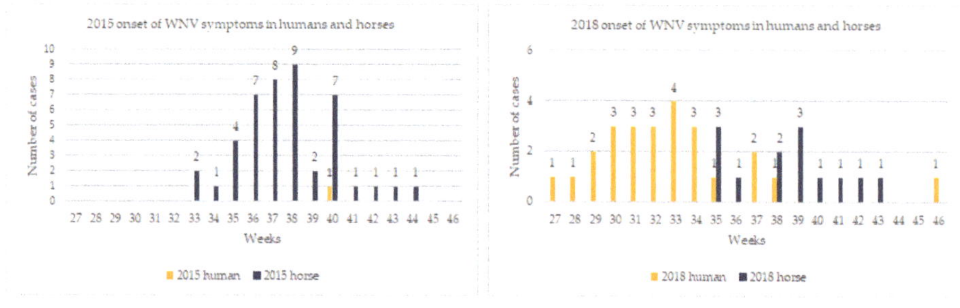


(**a**) (**b**)

**Figure 2.** (**a**) Weekly comparison of human and equine WNV cases notifications in 2015 in France and (**b**) weekly comparison of human and equine WNV cases notifications in 2018 in France. Diagrams depict dates of onset of symptoms whenever available, and in the absence of data on symptoms onset, date of veterinary samples (five cases in 2018) or sample reporting (one case in 2018) are shown.

Each of the 49 equine cases were identified in three departments surrounding the Camargue area, a large region delineated by the Rhone delta and characterized by high biological and environmental diversity in South-Eastern France. A total of 33 confirmed cases corresponding to 26 distinct outbreaks were located in Bouches–du-Rhône department, 15 confirmed cases (12 outbreaks) in Gard department, and one in Hérault department. Only one WNF human case was confirmed later in the season (2 October 2015, onset of symptoms 27 September) in Gard department and one mosquito pool corresponding to *Culex pipiens* mosquitoes was found positive in the same area on 11 September 2015 (Figure 3c,d).

In 2017, two human cases were diagnosed on the 21 August and 4 September in Alpes-Maritimes department. It was the first time in France that WNV was reported in this area. Following these cases, a serosurvey was carried out on 151 equids from a horse center located in the vicinity of the second WNV human case. Only one asymptomatic horse (1/151; CI 0%–2.9%) was found WNV-IgM positive (Figure 3e,f).

The WNV transmission season started much earlier and finished later in 2018 compared to the 2015–2017 period. Even if horses are particularly sensitive to the infection and can be used as indicators of virus circulation [32], autochthonous human cases were diagnosed before the occurrence of horse cases from 19 July 2018, while symptoms onset dated back to early July (week 27, 2–8 July). The number of human cases peaked on week 33 and the last case was notified week 46 (12–18 November) (Figure 2b). Four French departments around the Mediterranean Sea reported WNV infections (i.e., Alpes-Maritimes, Bouches-du-Rhône, Vaucluse, and Eastern Pyrenees). Alpes-Maritimes department was the first area reporting WNV cases and a cluster of 21 cases were located exclusively in this area (Figure 3h). Three raptors were found WNV positive later on in the season in the same area (in Nice and Antibes, Figure 3g). For the first time in 2018, an outbreak occurred in the south of Corsica with the first human case diagnosed at week 32 (6–12 August) and a second one at week 39 (24–30 September). Concomitantly, the onset of WNV equine outbreaks was reported on week 35 (starting 27 August) in North Corsica (Bastia) and four horses were found positive in 2018 in Corsica (Figure 3g). Finally, one long-eared owl (*Asio Otus*) was also diagnosed positive in South Corsica. Interestingly, the other 2018 equine cases were located in two departments in the Camargue area, namely Bouches-du-Rhône and Hérault, already affected by WNV equine outbreaks in 2015 (Figure 3g). The notification of WNV cases in equids started later than in humans and at a comparable period than in 2015 (week 33, 11 August 2015 versus week 35, 29 August 2018).

WNV activity in 2019 was lower than in 2018. It was characterized by a circulation of the virus in the Camargue region with WNV infected horses ($n$ = 11) and by the resurgence of WNV disease in Corsica (n = 2 horses). Moreover, one WNF and one WNND human cases were detected in Var, a department with sporadic WNV transmission to humans and horses identified since 2003 [33].

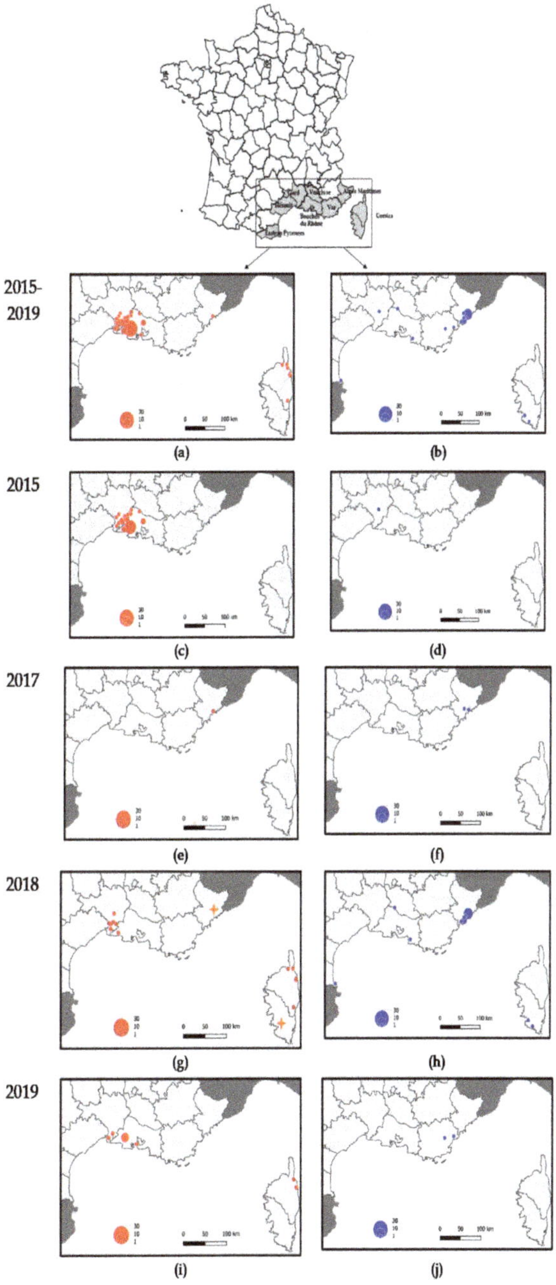


**Figure 3.** (**a**,**b**) Comparison of WNV case distribution in humans (blue dots) and horses (red dots) during 2015–2019; (**c**,**d**) Period 2015; (**e**,**f**) Period 2017; (**g**,**h**) Period 2018; and (**i**,**j**) Period 2019. In 2018, distribution of bird cases is represented by an orange star.

Recent changes in the temporal and spatial distribution of French WNV cases can be highlighted from 2015–2019 data analysis. Specifically, a multiplication of circulation foci have been reported during the last three years, with the emergence of WNV and recent description of clinical cases in the departments of Alpes-Maritimes, Var, and French Corsica island alongside the usual enzootic WNV circulation in the Camargue area during most of the period (2015, 2018–2019). WNV emergence in Alpes-Maritimes was associated with an increase of reported WNV cases in humans and birds but not in equids, while the distinct spatial distribution of human (mostly in Alpes-Maritimes) and horse (mostly in Camargue) WNV cases have been observed in Southern France these last years.

The intensity of WNV circulation and transmission is shaped in part by mosquito vector abundance and is influenced by biotic and abiotic factors favorable for mosquito proliferation [34,35]. Mosquito abundance was found to be significantly lower in Hérault than in the other two departments for the period 2015 to 2019 but the number of traps was six for this department compared to eight for the Bouches-du- Rhône and Gard (see Section 4.2 material and methods). The analysis of *Culex pipiens* mosquito abundance in the Camargue area (Bouches-Du-Rhône, Gard, and Hérault) during the vector season (June to October) indicated that the total number of trapped mosquitoes was significantly higher in 2018 than in 2016, 2017, and 2019 ($p$ values $\leq 0.002$). The difference observed between 2015 and 2018 was not found significant (Figure 4b and Table 1). Moreover, since the importance of WNV outbreaks in Europe was found to be strongly correlated with the length of the mosquito proliferation season, early abundance of mosquitoes in June was compared in 2015–2019. The vector season started earlier in 2018 than in other years, as the number of mosquitoes trapped in June was significantly higher in 2018 than in 2015, 2016, 2017, and 2019 ($p$ values $< 0.0007$) (Table 1). The abundance ratio for June 2015 was 0.20, thus corresponding to an abundance five times higher in June 2018 than in June 2015, after controlling for the effect of the department (Figure 4a, Table 1).

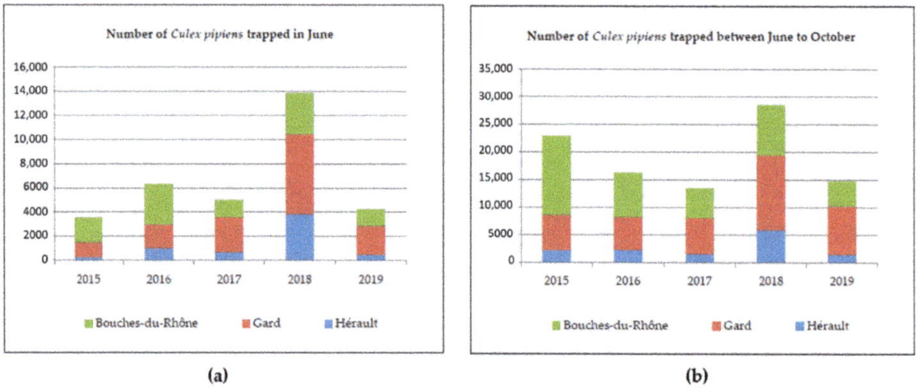


**Figure 4.** (**a**) Number of *Culex pipiens* mosquitoes trapped in three departments of the Camargue area (Bouches-Du-Rhône, Gard, and Hérault) in June during the 2015–2019 seasons and (**b**) number of *Culex pipiens* mosquitoes trapped in three departments of the Camargue area (Bouches-Du-Rhône, Gard and Hérault) between June to October during the 2015–2019 seasons.

**Table 1.** Negative binomial generalized linear models of the number of trapped *Culex pipiens* mosquitoes in three departments of Southern France, in June (a) or during the year (b), from 2015 to 2019.

| Variable | Value | *(a) Cx. pipiens* Trapped in June | | (b) Total *Cx. pipiens* Trapped | |
|---|---|---|---|---|---|
| | | Abundance Ratio | *p*-Value | Coefficient | *p*-Value |
| Department | Bouches du Rhône | Reference | | Reference | |
| | Gard | 1.14 | 0.53 | 1.02 | 0.92 |
| | Hérault | 0.36 | <0.0001 | 0.30 | <0.0001 |
| Year | 2018 | Reference | | Reference | |
| | 2015 | 0.20 | <0.0001 | 0.66 | 0.054 |
| | 2016 | 0.40 | 0.0007 | 0.51 | 0.002 |
| | 2017 | 0.30 | <0.0001 | 0.40 | <0.0001 |
| | 2019 | 0.24 | <0.0001 | 0.43 | <0.0001 |

### 2.2. First Description of WNV Lineage 2 Isolates in France in 2018

Phylogenetic analysis of the virus isolated from the brain of a WNV-infected horse in 2015 identified a lineage 1 strain belonging to the Western Mediterranean clade and genetically related to earlier French isolates collected in the Camargue area in 2000 and 2004 (Figure 5). It suggests an endemic circulation of the virus in the Camargue area, with WNV cycling in most years 2000–2014 between birds and *Culex* mosquitoes only and spilling over to horses and humans more regularly in 2015–2019.

Interestingly, WNV lineage 2 was recovered from raptor specimens found moribund in Alpes-Maritimes in 2018, demonstrating a recent emergence of WNV lineage 2 in Southeastern France. WNV strains showed the highest genetic homology with WNV strains reported recently in 2014 in Northern Italy (Veneto and Lumbardy) (Figure 5).

WNV strains detected in France, in one horse in 2015 (WNV-Akela/France/2015, indicated by a circle) and in wild birds in 2018 (WNV-6125/France/2018 and WNV-7025/France/2018, highlighted with triangles) belonging to different lineages, with a homology of 79.7–79.8% at the nucleotide level and 93.9–94.0%% (3223/3435) at the amino acid level (Table S1). 9–14, and eight amino acid substitutions affecting different viral genes were observed between WNV-Akela/France/2015 and older French lineage 1 isolates and between lineage 2 WNV-7025/France/2018, and the closely genetically related WNV-Cremona4/Italy/2014 respectively (Tables S2 and S3). No amino acid substitutions correspond to established WNV molecular virulence determinants or to positively selected codons [37,38].

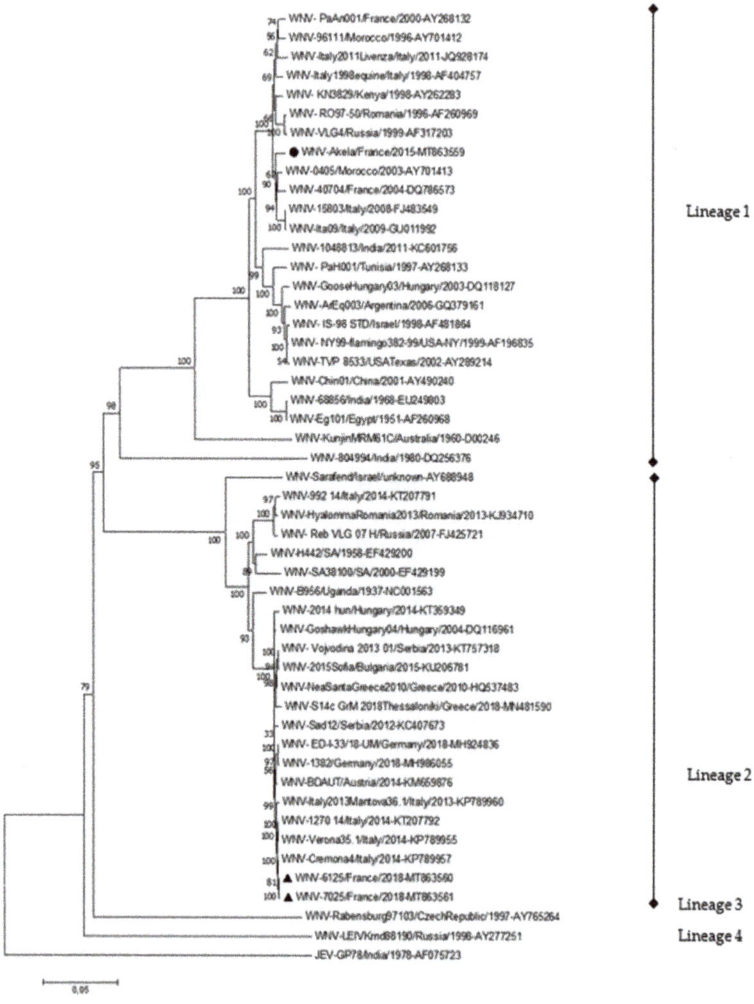


**Figure 5.** Molecular phylogenetic tree of WNV complete genome sequences detected in one horse (2015, black circle) and two birds (2018, black triangles) in France. The evolutionary history was inferred using the Neighbor–Joining and Maximum Likelihood methods in MEGA7 [36]. The optimal tree generated using the Neighbor–Joining method with the sum of branch length = 1,28228091 is shown. The percentage of replicate trees in which the associated taxa clustered together in the bootstrap test (1000 replicates) are shown next to the branches. The tree is drawn to scale, with branch lengths expressed in the same units as for the evolutionary distances used to infer the phylogenetic tree. The evolutionary distances were computed using the Jukes–Cantor method and are in the units of the number of base substitutions per site. The analysis involved 50 nucleotide sequences, including Japanese Encephalitis Virus as an outgroup. All positions containing gaps and missing data were eliminated. There were a total of 10,475 positions in the final dataset.

## 3. Discussion

In France, WNV caused outbreaks involving several human and horse cases in the beginning of the 1960s before it disappeared for 35 years [7]. During the 2000–2008 period, four episodes of WNV transmission were reported in France. Only WNV lineage 1 was reported in France during this period, with closely genetically related WNV isolates belonging to the Western Mediterranean clade identified in the Camargue area in 2000 (in horses) [39] and in 2004 (in birds) (Figure 5) [40]. It re-emerged following a cyclical and hardly predictable pattern and was mostly limited to Camargue, a high-risk area for WNV circulation due to high concentration of wetlands, mosquitoes, wild birds, and horses [41–43]. The same spatial Camargue location of equine outbreaks was pointed out in 2015, 2018, and 2019. The enhanced abundance of WNV competent mosquitoes in 2015 and 2018 and an earlier vector season in the Camargue area in 2018 have been identified as potential risk factors for higher WNV transmission. Moreover, the phylogenetic analysis of one WNV strain identified recently in Camargue supports WNV enzootic transmission in this region as it revealed that the sequence of WNV isolated from a confirmed equine case in 2015 is close to the lineage 1 strain that circulated in France in 2004 [44,45]. The percentage of nucleotide identity between French WNV lineage 1 isolates (>98.3%, Table S1) is coherent with the mean evolutionary rate of the European WNV strains ($3.7 \times 10^{-4}$ substitutions/site/year) [46]. Most of the mutations distinguishing the viral isolates were synonymous and homogenously distributed along the viral genome, which suggests that the genetic evolution of French WNV strains arose through a strong and local diversifying selection.

In 2018, WNV lineage 2 belonging to the Central and Eastern European clade (CEC) as defined by Ziegler et al. [47] was isolated for the first time from wild raptors, which have been shown to be particularly susceptible to WNV neuro-invasive infections, in Alpes-Maritimes in France [18,48,49]. WNV lineage 2 emergence in France was associated with exceptional WNV activity and lineage 2 spread in Western and Northern-most territories (Germany) in Europe this same year [25]. During this year, WNV infections in Europe increased dramatically compared to previous transmission seasons. From June to November 2018, a large part of Europe faced a period of unusually hot weather that led to record-breaking temperatures [50]. Like in France, European WNV infections started earlier in 2018 than in previous years. Indeed, the first WNF cases were reported on 31 May (week 22) in Greece which is the earliest disease onset compared with previous years [51]. At the end of 2018, a total of 1503 human infected confirmed cases were reported in 11 countries of the European Union (with almost 92% of cases coming from Italy, Greece, Romania, and Hungary) [52]. This number exceeded the cumulative number of WNV reported infections of the seven previous years [53]. The highest increase compared to previous transmission season was observed in Bulgaria (15 fold) followed by France (13.5 fold), and Italy (10.9 fold) [53]. During the 2018 transmission season, reports from the ECDC also indicated a high transmission among horses with 285 outbreaks reported by European member states as follows: 149 in Italy, 91 in Hungary, 15 in Greece, and 13 in France representing an increase of 30% in comparison with the number of outbreaks in 2017 [53].

In particular, in 2018, there was a large WNV lineage 2 outbreak in Northern Italy, including the Piemonte regions. WNV lineage 2 circulation was first documented in Italy in 2011 and, since then, has settled in Northern Italy at least since 2013 [14,54]. Recent phylogenetic analysis [55] revealed that two Italian lineage 2 strains, namely clade A and clade B diverged between 2010 and 2012 from a central region of the Po Valley. Clade A spread towards Northeastern Italy and apparently became extinct in 2013–2014, whereas clade B spread north-west reaching the most Western regions of Italy. Such a WNV short distance introduction via infected birds coming from a neighboring country has been usually hypothesized [18,47]. According to the high percentage of nucleotide homology (99.76%) between the Italian 2014 clade B and French 2018 WNV strains (Table S1), we hypothesize that WNV lineage 2 gradually spread from Northern Italy to South-Eastern France in 2018 or 2017, considering that WNV lineage 2 was isolated from wild birds in Alpes-Maritimes in 2018 and that human cases were already reported in the same area in 2017. Moreover, Histidine instead of Proline residues are found at position 249 in the helicase part of NS3 in recent Italian and French lineage 2 strains. The role of this genetic modification in the modulation of WNV pathogenicity in mammalian and bird hosts has been regularly debated [56,57]. While a specific non-synonymous substitution, Lys2114Arg, was identified in several WNV lineage 2 isolates obtained in Germany in 2018 and could be associated with an increased WNV fitness as such a mutation was not evidenced in the French 2018 lineage 2 strains.

In 2018, most human cases in mainland France were reported in areas with WNV-related bird mortality, which is consistent with the positive relationship between WND human cases and seroprevalence level in passerine birds, previously demonstrated at the European level [58]. The absence of horse cases before the onset of human cases could be explained by the very low density of horses in the Alpes-Maritimes area and by the location of 13 out of 26 human cases in the urban city of Nice [59]. The increase observed in 2018 in WNV human cases in France could result from differing virulence or transmission properties for humans and for horses of WNV lineages 1 and 2 and from varying animal and human densities in areas reporting WNV infections in 2018 (Alpes-Maritimes with densely populated urbanized areas, while most horse cases were reported in a natural wetland, the Camargue area); the first hypotheses (with more transmission to and more cases in humans associated with WNV lineage 2 infections) would deserve more attention but are currently not supported by the literature.

We also document in 2018 the detection of WNV for the first time in French Corsica Island. This finding is not surprising as a serosurvey carried out in Corsica in 2014 highlighted that 9.4% of horses presented WNV antibodies. Among these positive horses, 66.6% were native from the island, indicative of WNV local circulation [60]. The identification of WNV clinical cases in humans, horses, and birds in Corsica further documents recent and active circulation of the virus. Nevertheless, the identification of the causative lineage could not be achieved as a low viral load was evidenced in clinical specimens collected on the island (one raptor, a long eared-owl) and as both lineages 1 and 2 were described recently in Sardinia and Italy [61].

Fewer outbreaks were reported in 2019 than in 2018 in France and Europe, but a changing epidemiological pattern of WNV circulation can be anticipated in France in the coming years. Indeed, the introduction of WNV lineage 2 in Hungary in 2004 was followed by strain adaptation and limited activity in 2005–2007 while extensive spread of the virus was reported from 2008 [18]. Moreover, a remarkable extension of the distribution area of WNV lineage 2 has been evidenced in 2018. Specifically equine cases and mortality on resident wild and captive birds were detected for the first time in Eastern and Southeastern Germany [25]. This introduction was followed in Germany by an increase of equine WNV outbreaks in 2019 and the reporting of the first five confirmed mosquito-borne autochthonous human cases [47]. Another important finding during the 2018 transmission season relates to WNV genome detection in one blood donor for the first time in France by the French blood establishment [31]. Interestingly this donor had spent time in Alpes-Maritimes before the occurrence of the first animal or human WNV cases. These new data on WNV spread in South France and on positive WNV screenings in blood products and organ transplants highlight the necessity to strengthen WNV integrated surveillance in France in order to primarily secure human blood, cells, or organ products. WNV surveillance has been mainly supported by clinical surveillance programs focusing on the analysis of moribund or dead birds and of horses and human patients with neuroinvasive signs, which may lack sensitivity and fail to detect low-level circulation. A combination of clinical event-based surveillance activities and active monitoring of WNV enzootic transmission through the regular monitoring of seroconversions in sentinel and/or resident birds and horses or through mosquito trapping and WNV screening would enhance the chance to early detect WNV transmission [62].

Finally, an enhanced transmission of WNV in Southeastern France in 2018 paralleled an unusually high number of outbreaks of another *Culex*-borne flavivirus, Usutu, and in most French metropolitan territories [63]. Such findings emphasize the need of unraveling the virological, ecological, and climatic factors responsible for *Culex*-borne flavivirus emergence in France and Europe [64,65].

## 4. Materials and Methods

### 4.1. Samples

Main organizational aspects of the French West Nile virus surveillance system in animals, humans, and vectors have been described previously in the article of Bahuon et al. [11]. Briefly, the surveillance is based on clinical case definition (human and equine) or criteria for dead or sick birds' reports, collection, and testing. No routine indicator-based surveillance is implemented on the animal population. Diagnostic specimens are from 1/suspect human cases, as well as *Culex* mosquito populations sampled in affected areas once the viral circulation has been confirmed. They are analyzed by the National Reference center (NRC) for arboviruses (IRBA-Armed Forces Biomedical Research Institute) 2/each horse and avian suspect cases confirmed by the National Reference laboratory (NRL) on West Nile virus (Anses, Animal Health Laboratory, Maisons-Alfort) [12]. Suspect West Nile cases correspond to human patients over 15 years old and equids presenting with fever (≥38.5 °C) and symptoms of viral meningitis or encephalitis; wild or captive birds (raptors, corvids, and turdids more specifically) found dead, and individuals displaying neurological symptoms during the surveillance period (1 June to end of November) in the at-risk area (i.e., counties in the Mediterranean area). Moreover wild bird surveillance has been extended to departments considered, according to a statistical model, at an increased risk of WNV transmission, and located along the Mediterranean Sea and the Rhone River in South Eastern France, as well as in Bas-Rhin in North Eastern France since 2019 [58].

### 4.2. *Mosquito Collection*

Mosquitoes were collected weekly from mid-May to late October, corresponding to the mosquito season in the Rhône Delta, Camargue. CDC-like traps (John W. Hock Company, Gainesville, FL, USA) were used without light and were baited with carbon-dioxide dry ice (−80 °C). The trapping network was composed of 8 traps in the departments of Gard, 8 traps in Bouches-du-Rhône, and 6 traps in Hérault departments. Mosquitoes were stored in the fridge, killed, and identified with identification morphological keys.

### 4.3. *Serology*

Blood samples were collected in dry tubes, allowed to clot, and centrifuged at 1500 rpm for 10 min and stored at +4 °C during 1 month at most or at −20 °C for long term archiving.

For equine suspected cases reported to the French NRL, sera were first screened for anti-WNV antibodies by competition ELISA (ID Screen West Nile competition kit, IDVet Company, Montpellier, France) in local veterinary laboratories. Then IgG positive sera were further analyzed by M-antibody capture ELISA for IgM detection (ID screen West Nile IgM capture, IDVet company, Montpellier, France) in local veterinary laboratories and confirmed at the NRL. Analysis and interpretation of ELISAs were performed according to the manufacturer's instructions. In the event of IgM positive screening, the first samples collected during WNV outbreaks were confirmed by microneutralization test (MNT) as described in Beck et al. [66]. A confirmed case was therefore defined as a clinical suspected horse with at least a positive IgM ELISA test.

For human WNV diagnosis, sera and cerebrospinal fluid (CSF) were tested by in-house ELISAs (indirect IgG and MAC-ELISAs) using precipitated and inactivated virus. A case of WNV infection is confirmed with the presence of IgM in CSF and/or IgM and IgG in sera and anti-WNV neutralizing antibodies [67].

### 4.4. *Real-Time RT-PCR*

Brain of horses and birds, EDTA blood, and CSF suspected to be infected with WNV were stored at −80 °C until analysis. Brains were grinded in Dulbecco modified Eagle's minimal essential medium (DMEM) with ceramic beads (MP Biomedicals, Illkirch, France) and FastPrep ribolyzer in BSL3 facilities. A total of 560 μL of Lysis buffer from the QIAamp Viral RNA kit (Qiagen, Hilden, Germany) were added to 140 μL of grinded material before RNA extraction with the automate QIAcube. Human samples (EDTA blood and CSF) were processed the same way. Every RNA extracts were subjected to real time (rt) RT-PCR following the protocol described earlier by Linke et al. [68].

### 4.5. Virus Isolation

One milliliter of brain homogenates of WNV rtRT-PCR positive wild birds and horses was prepared in DMEM culture medium and inoculated on T25 flask that had been seeded with Vero NK cells (ATCC: CCL81™), 24 h earlier and washed with DMEM before inoculation. After 1h 30 of incubation at 37 °C with 5% $CO_2$, cells were washed twice with phosphate buffered saline (PBS), and complete medium (DMEM+ 1% penicillin- streptomycin+ 1% sodium pyruvate + 5% fetal calf serum) was added. The cells were observed each day from 3 days to 7 days post infection (pi). As soon as cytopathic effects (CPE) were detected, the supernatant was collected, stored at −80 °C, and RNA extracts subjected to rtRT-PCR to confirm WNV detection. Primary isolation was followed by a passage on Aedes albopictus (C6/36) (ATCC® CRL1660™) cell line. A total of 200µL–1 mL of Vero cell supernatants was added to T25 flask that had been seeded with C6/36 24 h earlier and washed with Leibowitz L15 medium before inoculation. After 1h 30 of incubation at 28 °C without $CO_2$, 6 mL of Leibowitz L15 media + 1% penicillin- streptomycin+ 1% sodium pyruvate + 1% L Glutamin+ 10% fetal calf serum were added. CPEs were not systematically observed in C6/36 cells and supernatants were collected on day 7 post-infection at the latest and tested as described above. This protocol is adapted from the OIE Manual of Diagnostic Tests and vaccines for Terrestrial Animals [69].

### 4.6. Nucleotide Sequencing and Sequence Analysis

Sequencing libraries were prepared from genomic RNAs extracted from virus isolate (2015) or from organ homogenates (2018) and whole-genome sequencing data were obtained as previously described (Ion Torrent sequencing and assembly with CLC Genomics Workbench for Genbank accession number MT863559, 2015 [70] or with bwa for Genbank accession numbers MT863560-1, 2018 [71]). Multiple alignment of the nucleotide sequences was performed using the ClustalW algorithm and phylogenetic analysis was performed using the Neighbor–Joining and Maximum Likelihood methods in MEGA7 [36].

### 4.7. Statistical Analysis

We used negative binomial generalized linear models to analyze the mosquito trapping data. The dependent variable was the number of trapped *Cx. pipiens*, and the independent variables were the department (Bouches-du-Rhône–reference class, Gard, or Hérault) and the year (2015–2019, the reference class being 2018). Two models were separately fitted: One for the yearly total number of trapped *Cx. pipiens*, and the other for the number of *Cx. pipiens* trapped in June. Statistical analyses were performed using R 3.6.1 [72].

**Supplementary Materials:** The following are available online at http://www.mdpi.com/2076-0817/9/11/908/s1. Table S1: Table presenting percentage homology in nucleotidic (nt, upper part) and in amino acid (aa, lower part) WNV sequences. Table S2: Amino acid substitutions identified between WNV-Akela/France/2015 and older French lineage 1 isolates. Table S3: Amino acid substitutions identified between lineage 2 WNV-7025/France/2018 and the closely genetically related WNV-Cremona4/Italy/2014.

**Author Contributions:** Conceptualization, C.B., S.D., I.L.G. and S.L. (Sylvie Lecollinet); methodology, C.B., I.L.G., S.D., B.D. and S.L. (Sylvie Lecollinet); software, B.D., Y.B., P.L. and X.d.L.; validation C.B., S.D., I.L.G. and S.L. (Sylvie Lecollinet), F.F., investigation, C.B., S.D., I.L.G., G.G. (Gaelle Gonzalez), M.D., S.L. (Steeve Lowenski), P.L., G.G. (Gilda Grard), G.A.D., J.T., G.L., and S.L. (Sylvie Lecollinet), F.F.; resources C.B., I.L.G., F.F., Y.B., P.L., X.d.L., G.L., S.D. and S.L. (Sylvie Lecollinet); data curation, C.B., I.L.G., F.F., G.G. (Gaelle Gonzalez), Y.B., P.L., X.d.L., G.G. (Gilda Grard), G.A.D., G.L., B.D. and S.L. (Steeve Lowenski ); writing—original draft preparation, C.B.; writing—review and editing, C.B., I.L.G., F.F., G.G. (Gaelle Gonzalez), S.Z., G.L., B.D., S.D. and S.L. (Sylvie Lecollinet); visualization, C.B.; supervision, C.B., G.G. (Gilda Grard), I.L.G. and S.L. (Sylvie Lecollinet). All authors have read and agreed to the published version of the manuscript.

**Funding:** This research was funded by EU Commission through DG SANTé (EU-RL on equine diseases).

**Acknowledgments:** The authors would like to thank the local veterinary laboratories, veterinary practitioners, local hunting federations, and Office français de la Biodiversité (OFB).

**Conflicts of Interest:** The authors declare no conflict of interest.

## References


1. Hubalek, Z.; Halouzka, J. West Nile fever—A reemerging mosquito-borne viral disease in Europe. *Emerg. Infect. Dis.* **1999**, *5*, 643–650. [CrossRef]
2. Kramer, L.D.; Li, J.; Shi, P.-Y. West Nile virus. *Lancet Neurol.* **2007**, *6*, 171–181. [CrossRef]
3. Perez-Ramirez, E.; Llorente, F.; del Amo, J.; Fall, G.; Sall, A.A.; Lubisi, A.; Lecollinet, S.; Vazquez, A.; Jimenez-Clavero, M.A. Pathogenicity evaluation of twelve West Nile virus strains belonging to four lineages from five continents in a mouse model: Discrimination between three pathogenicity categories. *J. Gen. Virol.* **2017**, *98*, 662–670. [CrossRef]
4. Beck, C.; Jimenez-Clavero, M.A.; Leblond, A.; Durand, B.; Nowotny, N.; Leparc-Goffart, I.; Zientara, S.; Jourdain, E.; Lecollinet, S. Flaviviruses in Europe: Complex circulation patterns and their consequences for the diagnosis and control of West Nile disease. *Int. J. Environ. Res. Public Health* **2013**, *10*, 6049–6083. [CrossRef]
5. Joubert, L.; Oudar, J.; Hannoun, C.; Beytout, D.; Corniou, B.; Guillon, J.C.; Panthier, R. Epidemiology of the West Nile virus: Study of a focus in Camargue. Iv. Meningo-Encephalomyelitis of the horse. *Ann. Inst. Pasteur* **1970**, *118*, 239–247.
6. Filipe, A.R.; Pinto, M.R. Survey for antibodies to arboviruses in serum of animals from Southern Portugal. *Am. J. Trop. Med. Hyg.* **1969**, *18*, 423–426. [CrossRef] [PubMed]
7. Murgue, B.; Murri, S.; Triki, H.; Deubel, V.; Zeller, H.G. West Nile in the Mediterranean basin: 1950–2000. *Ann. N. Y. Acad. Sci.* **2001**, *951*, 117–126. [CrossRef]
8. Sotelo, E.; Fernandez-Pinero, J.; Llorente, F.; Vazquez, A.; Moreno, A.; Aguero, M.; Cordioli, P.; Tenorio, A.; Jimenez-Clavero, M.A. Phylogenetic relationships of Western Mediterranean West Nile virus strains (1996–2010) using full-length genome sequences: Single or multiple introductions? *J. Gen. Virol.* **2001**, *92*, 2512–2522. [CrossRef] [PubMed]
9. Autorino, G.L.; Battisti, A.; Deubel, V.; Ferrari, G.; Forletta, R.; Giovannini, A.; Lelli, R.; Murri, S.; Scicluna, M.T. West Nile virus epidemic in horses, Tuscany Region, Italy. *Emerg. Infect. Dis.* **2002**, *8*, 1372–1378. [CrossRef] [PubMed]
10. Ceianu, C.S.; Ungureanu, A.; Nicolescu, G.; Cernescu, C.; Nitescu, L.; Tardei, G.; Petrescu, A.; Pitigoi, D.; Martin, D.; Ciulacu-Purcarea, V.; et al. West Nile virus surveillance in Romania: 1997–2000. *Viral Immunol.* **2001**, *14*, 251–262. [CrossRef]
11. Bahuon, C.; Marcillaud-Pitel, C.; Bournez, L.; Leblond, A.; Beck, C.; Hars, J.; Leparc-Goffart, I.; L'Ambert, G.; Paty, M.C.; Cavalerie, L.; et al. West Nile virus epizootics in the Camargue (France) in 2015 and reinforcement of surveillance and control networks. *Rev. Sci. Tech.* **2016**, *35*, 811–824. [CrossRef] [PubMed]
12. Johnson, N.; de Marco, M.F.; Giovannini, A.; Ippoliti, C.; Danzetta, M.L.; Svartz, G.; Erster, O.; Groschup, M.H.; Ziegler, U.; Mirazimi, A.; et al. Emerging mosquito-borne threats and the response from European and Eastern Mediterranean countries. *Int. J. Environ. Res. Public Health* **2018**, *15*, 2775. [CrossRef] [PubMed]
13. Del Giudice, P.; Schuffenecker, I.; Vandenbos, F.; Counillon, E.; Zellet, H. Human West Nile virus, France. *Emerg. Infect. Dis.* **2004**, *10*, 1885–1886. [CrossRef] [PubMed]
14. Lecollinet, S.; Pronost, S.; Coulpier, M.; Beck, C.; Gonzalez, G.; Leblond, A.; Tritz, P. Viral equine encephalitis, a growing threat to the horse population in Europe? *Viruses* **2019**, *12*, 23. [CrossRef]
15. de Heus, P.; Kolodziejek, J.; Camp, J.V.; Dimmel, K.; Bago, Z.; Hubalek, Z.; van den Hoven, R.; Cavalleri, J.V.; Nowotny, N. Emergence of West Nile virus lineage 2 in Europe: Characteristics of the first seven cases of West Nile neuroinvasive disease in horses in Austria. *Transbound. Emerg. Dis.* **2019**, *67*, 1189–1197. [CrossRef] [PubMed]
16. Hernández-Triana, L.M.; Jeffries, C.L.; Mansfield, K.L.; Carnell, G.; Fooks, A.R.; Johnson, N. Emergence of West Nile virus lineage 2 in Europe: A review on the introduction and spread of a mosquito-borne disease. *Front. Public Health* **2014**, *2*, 271. [CrossRef]
17. Bakonyi, T.; Ivanics, E.; Erdelyi, K.; Ursu, K.; Ferenczi, E.; Weissenbock, H.; Nowotny, N. Lineage 1 and 2 strains of encephalitic West Nile virus, Central Europe. *Emerg. Infect. Dis.* **2006**, *12*, 618–623. [CrossRef]
18. Bakonyi, T.; Ferenczi, E.; Erdelyi, K.; Kutasi, O.; Csorgo, T.; Seidel, B.; Weissenbock, H.; Brugger, K.; Ban, E.; Nowotny, N. Explosive spread of a neuroinvasive lineage 2 West Nile virus in Central Europe, 2008/2009. *Vet. Microbiol.* **2013**, *165*, 61–70. [CrossRef]

19. Papa, A.; Danis, K.; Baka, A.; Bakas, A.; Dougas, G.; Lytras, T.; Theocharopoulos, G.; Chrysagis, D.; Vassiliadou, E.; Kamaria, F.; et al. Ongoing outbreak of West Nile virus infections in humans in Greece, July–August 2010. *Eurosurveillance* **2010**, *15*, 19644. [CrossRef]
20. Napp, S.; Petric, D.; Busquets, N. West Nile virus and other mosquito-borne viruses present in Eastern Europe. *Pathog. Glob. Health* **2018**, *112*, 233–248. [CrossRef]
21. Merdic, E.; Peric, L.; Pandak, N.; Kurolt, I.C.; Turic, N.; Vignjevic, G.; Stolfa, I.; Milas, J.; Bogojevic, M.S.; Markotic, A. West Nile virus outbreak in humans in Croatia, 2012. *Coll. Antropol.* **2013**, *37*, 943–947. [PubMed]
22. Savini, G.; Capelli, G.; Monaco, F.; Polci, A.; Russo, F.; di Gennaro, A.; Marini, V.; Teodori, L.; Montarsi, F.; Pinoni, C.; et al. Evidence of West Nile virus lineage 2 circulation in Northern Italy. *Vet. Microbiol.* **2012**, *158*, 267–273. [CrossRef] [PubMed]
23. Busquets, N.; Laranjo-Gonzalez, M.; Soler, M.; Nicolas, O.; Rivas, R.; Talavera, S.; Villalba, R.; Miguel, E.S.; Torner, N.; Aranda, C.; et al. Detection of West Nile virus lineage 2 in North-Eastern Spain (Catalonia). *Transbound. Emerg. Dis.* **2019**, *66*, 617–621. [CrossRef] [PubMed]
24. Michel, F.; Sieg, M.; Fischer, D.; Keller, M.; Eiden, M.; Reuschel, M.; Schmidt, V.; Schwehn, R.; Rinder, M.; Urbaniak, S.; et al. Evidence for West Nile virus and Usutu virus infections in wild and resident birds in Germany, 2017 and 2018. *Viruses* **2019**, *11*, 674. [CrossRef]
25. Ziegler, U.; Luhken, R.; Keller, M.; Cadar, D.; van der Grinten, E.; Michel, F.; Albrecht, K.; Eiden, M.; Rinder, M.; Lachmann, L.; et al. West Nile virus epizootic in Germany, 2018. *Antiviral. Res.* **2019**, *162*, 39–43. [CrossRef]
26. Platonov, A.E.; Karan, L.S.; Shopenskaia, T.A.; Fedorova, M.V.; Koliasnikova, N.M.; Rusakova, N.M.; Shishkina, L.V.; Arshba, T.E.; Zhuravlev, V.I.; Govorukhina, M.V.; et al. Genotyping of West Nile fever virus strains circulating in Southern Russia as an epidemiological investigation method: Principles and Results. *Zhurnal Mikrobiol. Epidemiol. Immunobiol.* **2011**, *2*, 29–37.
27. Cotar, A.I.; Falcuta, E.; Dinu, S.; Necula, A.; Birlutiu, V.; Ceianu, C.S.; Prioteasa, F.L. West Nile virus lineage 2 in Romania, 2015–2016: Co-circulation and strain replacement. *Parasit. Vectors* **2018**, *11*, 562. [CrossRef]
28. Ravagnan, S.; Montarsi, F.; Cazzin, S.; Porcellato, E.; Russo, F.; Palei, M.; Monne, I.; Savini, G.; Marangon, S.; Barzon, L.; et al. First report outside eastern europe of West Nile virus lineage 2 related to the Volgograd 2007 Strain, Northeastern Italy, 2014. *Parasit. Vectors* **2015**, *8*, 418. [CrossRef]
29. Papa, A.; Papadopoulou, E.; Chatzixanthouliou, C.; Glouftsios, P.; Pappa, S.; Pervanidou, D.; Georgiou, L. Emergence of West Nile virus lineage 2 belonging to the Eastern European Subclade, Greece. *Arch. Virol.* **2019**, *164*, 1673–1675. [CrossRef]
30. Saegerman, C.; Alba-Casals, A.; Garcia-Bocanegra, I.; Pozzo, F.D.; van Galen, G. Clinical sentinel surveillance of equine West Nile fever, Spain. *Transbound. Emerg. Dis.* **2016**, *63*, 184–193. [CrossRef]
31. Ramalli, L.; Grard, G.; Beck, C.; Gallian, P.; L'Ambert, G.; Desvaux, S.; Jourdan, M.; Ortmans, C.; Paty, M.C.; Franke, F. West Nile virus infections in France, July to November 2018. *Eur. J. Public Health* **2019**, *29*, ckz186.631. [CrossRef]
32. Leblond, A.; Hendrikx, P.; Sabatier, P. West Nile virus outbreak detection using syndromic monitoring in horses. *Vector Borne Zoonotic Dis.* **2007**, *7*, 403–410. [CrossRef] [PubMed]
33. Durand, B.; Dauphin, G.; Zeller, H.; Labie, J.; Schuffenecker, I.; Murri, S.; Moutou, F.; Zientara, S. Serosurvey for West Nile virus in horses in Southern France. *Vet. Rec.* **2005**, *157*, 711–713. [CrossRef] [PubMed]
34. Paz, S.; Semenza, J.C. Environmental drivers of West Nile fever epidemiology in Europe and Western Asia—A review. *Int. J. Environ. Res. Public Health* **2013**, *10*, 3543–3562. [CrossRef] [PubMed]
35. Calzolari, M.; Angelini, P.; Bolzoni, L.; Bonilauri, P.; Cagarelli, R.; Canziani, S.; Cereda, D.; Cerioli, M.P.; Chiari, M.; Galletti, G.; et al. Enhanced West Nile virus circulation in the Emilia-Romagna and Lombardy Regions (Northern Italy) in 2018 detected by entomological surveillance. *Front. Vet. Sci.* **2020**, *7*, 243. [CrossRef]
36. Kumar, S.; Stecher, G.; Tamura, K. Mega7: Molecular evolutionary genetics analysis version 7.0 for bigger datasets. *Mol. Biol. Evol.* **2016**, *33*, 1870–1874. [CrossRef]
37. Chaintoutis, S.C.; Papa, A.; Pervanidou, D.; Dovas, C.I. Evolutionary dynamics of lineage 2 West Nile virus in Europe, 2004–2018: Phylogeny, selection pressure and phylogeography. *Mol. Phylogenet. Evol.* **2019**, *141*, 106617. [CrossRef]
38. Kaiser, J.A.; Wang, T.; Barrett, A.D. Virulence determinants of West Nile virus: How can these be used for vaccine design? *Future Virol.* **2019**, *12*, 283–295. [CrossRef]

39. Charrel, R.N.; Brault, A.C.; Gallian, P.; Lemasson, J.J.; Murgue, B.; Murri, S.; Pastorino, B.; Zeller, H.; de Chesse, R.; de Micco, P.; et al. Evolutionary relationship between old world West Nile virus strains. evidence for viral gene flow between Africa, the Middle East, and Europe. *Virology* **2003**, *315*, 381–388. [CrossRef]
40. Jourdain, E.; Schuffenecker, I.; Korimbocus, J.; Reynard, S.; Murri, S.; Kayser, Y.; Gauthier-Clerc, M.; Sabatier, P.; Zeller, H.G. West Nile virus in wild resident birds, Southern France, 2004. *Vector Borne Zoonotic Dis.* **2007**, *7*, 448–452. [CrossRef]
41. Bahuon, C.; Pitel, C.M.; Bournez, L.; Leblond, A.; Hars, J.; Beck, C.; Goffart, I.L.; L'Ambert, G.; Paty, M.C.; Cavalerie, L.; et al. Wnv epizootics in Camargue, France, 2015 and reinforcement of Wnv surveillance and control networks. *OIE Bull. Épidémiol.* **2016**. [CrossRef]
42. Murgue, B.; Murri, S.; Zientara, S.; Durand, B.; Durand, J.P.; Zeller, H. West Nile outbreak in horses in Southern France, 2000: The return after 35 years. *Emerg. Infect. Dis.* **2001**, *7*, 692–696. [CrossRef]
43. Pradier, S.; Sandoz, A.; Paul, M.C.; Lefebvre, G.; Tran, A.; Maingault, J.; Lecollinet, S.; Leblond, A. Importance of wetlands management for West Nile virus circulation risk, Camargue, Southern France. *Int. J. Environ. Res. Public Health* **2014**, *11*, 7740–7754. [CrossRef] [PubMed]
44. Vittecoq, M.; Lecollinet, S.; Jourdain, E.; Thomas, F.; Blanchon, T.; Arnal, A.; Lowenski, S.; Gauthier-Clerc, M. Recent circulation of West Nile virus and potentially other closely related flaviviruses in Southern France. *Vector Borne Zoonotic Dis.* **2013**, *13*, 610–613. [CrossRef]
45. Balanca, G.; Gaidet, N.; Savini, G.; Vollot, B.; Foucart, A.; Reiter, P.; Boutonnier, A.; Lelli, R.; Monicat, F. Low West Nile virus circulation in wild birds in an area of recurring outbreaks in Southern France. *Vector Borne Zoonotic Dis.* **2009**, *9*, 737–741. [CrossRef]
46. McMullen, A.R.; Albayrak, H.; May, F.J.; Davis, C.T.; Beasley, D.W.C.; Barrett, A.D.T. Molecular evolution of lineage 2 West Nile virus. *J. Gen. Virol.* **2013**, *94*, 318–325. [CrossRef] [PubMed]
47. Ziegler, U.; Santos, P.D.; Groschup, M.H.; Hattendorf, C.; Eiden, M.; Hoper, D.; Eisermann, P.; Keller, M.; Michel, F.; Klopfleisch, R.; et al. West Nile virus epidemic in Germany triggered by epizootic emergence, 2019. *Viruses* **2020**, *12*, 448. [CrossRef] [PubMed]
48. Perez-Ramirez, E.; Llorente, F.; Jimenez-Clavero, M.A. Experimental infections of wild birds with West Nile virus. *Viruses* **2014**, *6*, 752–781. [CrossRef]
49. Vidaña, B.; Busquets, N.; Napp, S.; Pérez-Ramírez, E.; Jiménez-Clavero, M.; Johnson, N. The role of birds of prey in West Nile virus epidemiology. *Vaccines* **2020**, *8*, 550. [CrossRef]
50. Zana, B.; Erdelyi, K.; Nagy, A.; Mezei, E.; Nagy, O.; Takacs, M.; Bakonyi, T.; Forgach, P.; Korbacska-Kutasi, O.; Feher, O.; et al. Multi-approach investigation regarding the West Nile virus situation in Hungary, 2018. *Viruses* **2020**, *12*, 123. [CrossRef]
51. Haussig, J.M.; Young, J.J.; Gossner, C.M.; Mezei, E.; Bella, A.; Sirbu, A.; Pervanidou, D.; Drakulovic, M.B.; Sudre, B. Early start of the West Nile fever transmission season 2018 in Europe. *Eurosurveillance* **2018**, *23*, 1800428. [CrossRef] [PubMed]
52. ECDC. West Nile Virus Infections by Affected Areas, in the EU/EEA Member States and EU Neighbouring Countries, 2018 Transmission Season. Available online: https://www.ecdc.europa.eu/en/publications-data/west-nile-virus-infections-affected-areas-eueea-member-states-and-eu-neighbouring (accessed on 14 May 2020).
53. ECDC. Epidemiological Update: West Nile Virus Transmission Season in Europe. 2018. Available online: https://www.ecdc.europa.eu/en/news-events/epidemiological-update-west-nile-virus-transmission-season-europe-2018 (accessed on 14 December 2018).
54. Marini, G.; Calzolari, M.; Angelini, P.; Bellini, R.; Bellini, S.; Bolzoni, L.; Torri, D.; Defilippo, F.; Dorigatti, I.; Nikolay, B.; et al. A quantitative comparison of West Nile virus incidence from 2013 to 2018 in Emilia-Romagna, Italy. *PLoS Negl. Trop. Dis.* **2020**, *14*, e0007953. [CrossRef] [PubMed]
55. Veo, C.; della Ventura, C.; Moreno, A.; Rovida, F.; Percivalle, E.; Canziani, S.; Torri, D.; Calzolari, M.; Baldanti, F.; Galli, M.; et al. Evolutionary dynamics of the lineage 2 West Nile virus that caused the largest European epidemic: Italy 2011–2018. *Viruses* **2019**, *11*, 814. [CrossRef]
56. Langevin, S.A.; Bowen, R.A.; Reisen, W.K.; Andrade, C.C.; Ramey, W.N.; Maharaj, P.D.; Anishchenko, M.; Kenney, J.L.; Duggal, N.K.; Romo, H.; et al. Host competence and helicase activity differences exhibited by West Nile viral variants expressing Ns3-249 amino acid polymorphisms. *PLoS ONE* **2014**, *9*, e100802. [CrossRef]

57. Dridi, M.; van den Berg, T.; Lecollinet, S.; Lambrecht, B. Evaluation of the pathogenicity of West Nile virus (WNV) lineage 2 strains in a SPF chicken model of infection: Ns3-249Pro mutation is neither sufficient nor necessary for conferring virulence. *Vet. Res.* **2015**, *46*, 130. [CrossRef]
58. Durand, B.; Tran, A.; Balanca, G.; Chevalier, V. Geographic variations of the bird-borne structural risk of West Nile virus circulation in Europe. *PLoS ONE* **2017**, *12*, e0185962. [CrossRef]
59. IFCE. Tableau Économique, Statistique et Graphique du Cheval en France Données 2016/2017. 2017. Available online: https://www.ifce.fr/wp-content/uploads/2017/11/OESC-Annuaire-ECUS-2017.pdf (accessed on 14 May 2020).
60. Maquart, M.; Dahmani, M.; Marie, J.L.; Gravier, P.; Leparc-Goffart, I.; Davoust, B. First serological evidence of West Nile virus in horses and dogs from Corsica Island, France. *Vector Borne Zoonotic Dis.* **2017**, *17*, 275–277. [CrossRef]
61. Magurano, F.; Remoli, M.E.; Baggieri, M.; Fortuna, C.; Marchi, A.; Fiorentini, C.; Bucci, P.; Benedetti, E.; Ciufolini, M.G.; Rizzo, C.; et al. Circulation of West Nile virus lineage 1 and 2 during an outbreak in Italy. *Clin. Microbiol. Infect.* **2012**, *18*, E545–E547. [CrossRef] [PubMed]
62. Rizzoli, A.; Jimenez-Clavero, M.A.; Barzon, L.; Cordioli, P.; Figuerola, J.; Koraka, P.; Martina, B.; Moreno, A.; Nowotny, N.; Pardigon, N.; et al. The challenge of West Nile virus in Europe: Knowledge gaps and research priorities. *Eurosurveillance* **2015**, *20*, 21135. [CrossRef]
63. Beck, C.; Gonzalez, G.; Decors, A.; Lemberger, K.; Lowenski, S.; Dumarest, D.; Lecollinet, S. Surveillance épidémiologique du virus Usutu dans l'avifaune. *Virologie* **2018**, *22*, 261–263.
64. Semenza, J.C.; Tran, A.; Espinosa, L.; Sudre, B.; Domanovic, D.; Paz, S. Climate change projections of West Nile virus infections in Europe: Implications for blood safety practices. *Environ. Health* **2016**, *15*, 28. [CrossRef]
65. Chevalier, V.; Tran, A.; Durand, B. Predictive modeling of West Nile virus transmission risk in the Mediterranean basin: How far from landing? *Int. J. Environ. Res. Public Health* **2013**, *11*, 67–90. [CrossRef] [PubMed]
66. Beck, C.; Despres, P.; Paulous, S.; Vanhomwegen, J.; Lowenski, S.; Nowotny, N.; Durand, B.; Garnier, A.; Blaise-Boisseau, S.; Guitton, E.; et al. A high-performance multiplex immunoassay for serodiagnosis of flavivirus-associated neurological diseases in horses. *BioMed Res. Int.* **2015**, *2015*, 678084. [CrossRef] [PubMed]
67. Carteaux, G.; Maquart, M.; Bedet, A.; Contou, D.; Brugières, P.; Fourati, S.; de Langavant, L.C.; de Broucker, T.; Brun-Buisson, C.; Leparc-Goffart, I.; et al. Zika virus associated with meningoencephalitis. *New Engl. J. Med.* **2016**, *374*, 1595–1596. [CrossRef] [PubMed]
68. Linke, S.; Ellerbrok, H.; Niedrig, M.; Nitsche, A.; Pauli, G. Detection of West Nile virus lineages 1 and 2 by real-time PCR. *J. Virol. Methods* **2007**, *146*, 355–358. [CrossRef]
69. OIE. Manual of Diagnostic Tests and Vaccines for Terrestrial Animals. 2018. Available online: https://www.oie.int/fileadmin/Home/eng/Health_standards/tahm/3.01.24_WEST_NILE.pdf (accessed on 14 May 2020).
70. Piorkowski, G.; Richard, P.; Baronti, C.; Gallian, P.; Charrel, R.; Leparc-Goffart, I.; de Lamballerie, X. Complete coding sequence of Zika virus from Martinique outbreak in 2015. *New Microbes New Infect.* **2016**, *11*, 52–53. [CrossRef]
71. Sailleau, C.; Breard, E.; Viarouge, C.; Gorlier, A.; Quenault, H.; Hirchaud, E.; Touzain, F.; Blanchard, Y.; Vitour, D.; Zientara, S. Complete genome sequence of bluetongue virus serotype 4 that emerged on the French island of Corsica in December 2016. *Transbound. Emerg. Dis.* **2018**, *65*, e194–e197. [CrossRef]
72. R Core Team. *R: A Language and Environment for Statistical Computing*; R Foundation for Statistical Computing: Vienna, Austria, 2013; Available online: http://www.R-project.org (accessed on 14 May 2020).


**Publisher's Note:** MDPI stays neutral with regard to jurisdictional claims in published maps and institutional affiliations.

*pathogens*

*Article*

# West Nile Virus Seroprevalence in the Italian Tuscany Region from 2016 to 2019

**Serena Marchi [1,*], Emanuele Montomoli [1,2], Simonetta Viviani [1], Simone Giannecchini [3], Maria A. Stincarelli [3], Gianvito Lanave [4], Michele Camero [4], Caterina Alessio [1], Rosa Coluccio [1,2] and Claudia Maria Trombetta [1]**

[1] Department of Molecular and Developmental Medicine, University of Siena, 53100 Siena, Italy; emanuele.montomoli@unisi.it (E.M.); simonetta.viviani@unisi.it (S.V.); caterina.alessio@student.unisi.it (C.A.); rosa.coluccio@gmail.com (R.C.); trombetta@unisi.it (C.M.T.)
[2] VisMederi S.r.l., 53100 Siena, Italy
[3] Department of Experimental and Clinical Medicine, University of Florence, 50134 Firenze, Italy; simone.giannecchini@unifi.it (S.G.); mariastincarelli@gmail.com (M.A.S.)
[4] Department of Veterinary Medicine, University of Bari, 70010 Valenzano, Italy; gianvito.lanave@gmail.com (G.L.); michele.camero@uniba.it (M.C.)
* Correspondence: serena.marchi2@unisi.it

**Citation:** Marchi, S.; Montomoli, E.; Viviani, S.; Giannecchini, S.; Stincarelli, M.A.; Lanave, G.; Camero, M.; Alessio, C.; Coluccio, R.; Trombetta, C.M. West Nile Virus Seroprevalence in the Italian Tuscany Region from 2016 to 2019. *Pathogens* **2021**, *10*, 844. https://doi.org/10.3390/pathogens10070844

Academic Editor: Francisco Llorente

Received: 28 April 2021
Accepted: 2 July 2021
Published: 5 July 2021



**Abstract:** Although in humans West Nile virus is mainly the cause of mild or sub-clinical infections, in some cases a neuroinvasive disease may occur predominantly in the elderly. In Italy, several cases of West Nile virus infection are reported every year. Tuscany was the first Italian region where the virus was identified; however, to date only two cases of infection have been reported in humans. This study aimed at evaluating the prevalence of antibodies against West Nile virus in the area of Siena Province to estimate the recent circulation of the virus. Human serum samples collected in Siena between 2016 and 2019 were tested for the presence of antibodies against West Nile virus by ELISA. ELISA positive samples were further evaluated using immunofluorescence, micro neutralization, and plaque reduction neutralization assays. In total, 1.9% (95% CI 1.2–3.1) and 1.4% (95% CI 0.8–2.4) of samples collected in 2016–2017 were positive by ELISA and immunofluorescence assay, respectively. Neutralizing antibodies were found in 0.7% (95% CI 0.3–1.5) of samples. Additionally, 0.9% (95% CI 0.4–1.7) and 0.65% (95% CI 0.3–1.45) of samples collected in 2018–2019 were positive by ELISA and immunofluorescence assay, respectively. The prevalence of neutralizing antibodies was 0.5% (95% CI 0.2–1.3). Although no human cases of West Nile infection were reported in the area between 2016 and 2019 and virus prevalence in the area of Siena Province was as low as less than 1%, the active asymptomatic circulation confirms the potential concern of this emergent virus for human health.

**Keywords:** West Nile virus; antibody; seroprevalence; Italy

## 1. Introduction

West Nile virus (WNV) is an emerging virus involving birds and Culex mosquitoes in its transmission cycle. Spillover events from this cycle involve mammalian hosts, in particular horses and humans, considered dead-end hosts [1]. In humans, although most infections (about 80%) are asymptomatic, those with a clinical manifestation present with a mild febrile illness, known as West Nile fever, are often underdiagnosed. In some cases, especially among the elderly, a more severe infection may develop as West Nile neuroinvasive disease (WNND), associated with significant morbidity and mortality [2]. As WNND is observed in less than 1% of infected subjects, the frequency of subclinical infections leads to an underestimation of the actual circulation of the virus.

In Italy, WNV is endemic in the Northern regions, where several cases of infection in humans are reported every year. However, in recent years the virus has expanded its distribution with cases also registered in Central and Southern Italy [3].

WNV was reported for the first time in Italy in 1998 among horses residing in wetland areas of Tuscany, Central Italy [4], while the first human case of WNND was detected ten years later, in 2008, in the Emilia-Romagna region, Northern Italy [5]. However, a retrospective study showed that in 2007 a woman living in Tuscany was infected with WNV [6], demonstrating that the virus was already circulating among humans in Central Italy before its isolation in Emilia-Romagna causing unrecognized human disease.

Following the identification of the first human cases of WNV infection, specific WNND surveillance systems were set up in the Emilia-Romagna and Veneto regions [7], followed by the implementation of national veterinary and human surveillance plans [8]. Surveillance activities include entomological, veterinary and human surveillance to be carried out from June to November, identified as the high-risk transmission period. Since 2008, WNV circulation has been reported in 10 Italian regions (Emilia-Romagna, Veneto, Lombardy, Sardinia, Sicily, Friuli Venezia Giulia, Piedmont, Tuscany, Basilicata, Apulia). From 2008 to 2017 a total of 231 cases of human WNND were reported [9].

The presence of WNV in Tuscany was reported in horses in 1998 [4], in 2009 [10], and in 2016 [11]. Although Tuscany was the first Italian region where the presence of WNV human infections was identified already in 2007 [6], since then only one imported case in 2011 in the province of Pisa [12] and two WNND cases in 2017 in the province of Livorno, a coastal area of the region, were reported [13] (Figure 1).

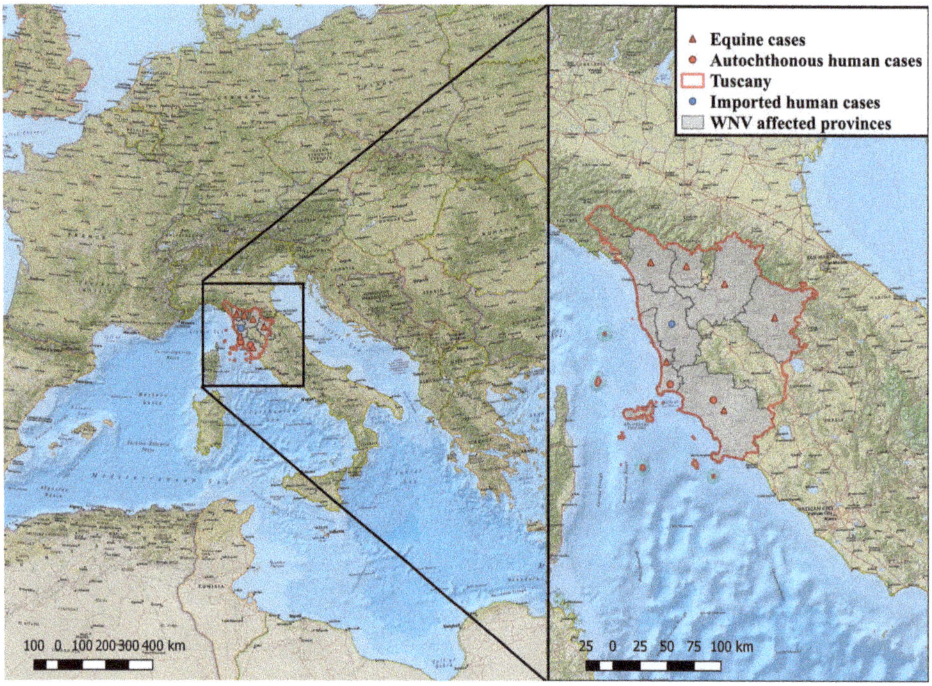


**Figure 1.** West Nile disease equine and autochthonous and imported human reported cases retrieved in Tuscany since 1998. The territories affected with West Nile virus (WNV) are colored in grey and those that are WNV-free are colorless. The maps were generated with Esri ArcGis Desktop 10.6.1 (www.esri.com, accessed on 2 June 2021). Red triangles, equine cases; Red circles, autochthonous human cases; Blue circle, imported human cases.

In Italy, human cases of WNV infection are usually detected starting from July and peaking in August–September. However, in 2018 the transmission season started earlier with the first detection of WNV from a pool of Culex mosquitoes in the Veneto region on the 7th of June [14], and the first confirmed human case was reported just 9 days later in the same province [15]. As of December 2018, a total of 577 confirmed cases of human infection were reported in the Veneto, Emilia-Romagna, Lombardy, Piedmont, Sardinia, and Friuli Venezia Giulia regions. In that very year, veterinary surveillance reported an increase in the circulation of WNV in mosquitoes, birds and horses in nine Italian regions (Emilia-Romagna, Veneto, Lombardy, Sardinia, Friuli Venezia Giulia, Piedmont, Lazio, Basilicata, and Apulia) [16].

Since 2018, the coastal provinces of Tuscany have been considered endemic areas [17], and since 2019, the province of Siena has been included among areas at high risk of transmission by national surveillance plans [18,19]. However, to date no cases of infection have been reported by routine surveillance activities.

Reports of epidemics in equine holdings suggest the circulation of WNV in the Tuscany region. However, human cases reported by routine surveillance are few. To date, limited data are available on the prevalence of WNV in the region.

The primary aim of this study was to evaluate the prevalence of WNV antibodies in the Siena Province area, in the Tuscany region, to estimate the recent circulation of WNV in an area where no infection cases have ever been reported by surveillance activities. The second aim was to investigate any variation in WNV prevalence in the province after the increase in transmission observed during the 2018 season.

## 2. Results

A total of 1800 samples, 879 for the years 2016–2017 and 921 for the years 2018–2019, were tested. The median age of all subjects whose serum samples were included in the study was 51 years, 50 years (age range 20–93 years) and 51 years (age range 20–94 years) for those sampled in 2016–2017 and in 2018–2019, respectively.

Collected samples were stratified by year of collection (2016–2017 and 2018–2019) and further stratified by sex and age group (20–60 and >60 years old). IgG borderline and positive samples identified by ELISA, immunofluorescence assay (IFA) and micro-neutralization (MN)/plaque reduction neutralization (PRN) assays in different years of collection by age group are reported in Table 1.

Out of the 879 samples collected in 2016–2017, 17 (1.9%, 95% confidence interval (CI) 1.2–3.1) samples were ELISA IgG positive or borderline. Comparisons of ELISA IgG positive results with sex ($p = 0.83$) and age groups ($p = 0.69$) did not show any statistical significance. Twelve samples (1.4%, 95% CI 0.8–2.4) were confirmed IgG positive by IFA. IFA IgG positive results were also not statistically associated with sex ($p = 0.85$) and age groups ($p = 0.71$). Six samples were positive by MN and PRN assays, thus showing a total prevalence of 0.7% (95% CI 0.3–1.5) of samples with neutralizing antibodies. MN/PRN positive results lack any statistical significance with sex ($p = 0.87$) and age groups ($p = 0.74$).

Out of the 879 samples collected in 2016–2017, 92 were also tested by ELISA IgM. Four samples were found positive (4.35%, 95% CI 1.36–11.0), one of which was also positive for ELISA IgG, IFA IgG, and neutralizing antibodies.

Out of the 921 samples collected in 2018–2019, eight (0.9%, 95% CI 0.4–1.7) samples were positive by ELISA IgG. Comparisons of ELISA IgG positive results with sex ($p = 0.18$) and age groups ($p = 0.71$) did not yield any statistical significance. Six samples (0.65%, 95% CI 0.3–1.45) were confirmed IgG positive by IFA. IFA IgG positive results were also not statistically associated with sex ($p = 0.23$) and age group ($p = 0.77$). Five samples were positive by MN and PRN for a total prevalence of 0.5% (95% CI 0.2–1.3). MN/PRN positive results lack any statistical significance by comparison with sex ($p = 0.26$) and age groups ($p = 0.79$).

Table 2 shows a summary of results with the characteristics of subjects who showed neutralizing antibodies to WNV. The median age was 57 years (age range 30–91 years).

In the univariate logistic regression model, the independent variables, sex and age group, did not show, consistently, statistically significant associations with IgG positive and borderline results for both 2016–2017 and 2018–2019 years of collection. In the multivariate logistic regression model, the independent variables confirmed the lack of association with IgG positive or borderline results (Table 3).

**Table 1.** ELISA IgG, IFA IgG, and MN/PRN borderline and positive samples by years of collection and age group. Data on sex and age group of collected and positive samples were summarized as counts and percentages and 95% confidence intervals were calculated.

| Study Population | 2016–2017 | | | | | | 2018–2019 | | | | | |
|---|---|---|---|---|---|---|---|---|---|---|---|---|
| | M | F | Total | ELISA IgG | IFA IgG | MN/PRN | M | F | Total | ELISA IgG | IFA IgG | MN/PRN |
| 20–60 years | 255 (43.8) | 327 (56.2) | 582 (66.2) | 10 (1.7%, 0.9–3.2) | 8 (1.4%, 0.65–2.7) | 3 (0.5%, 0.1–1.6) | 307 (50.1) | 306 (49.9) | 613 (66.6) | 6 (1.0%, 0.4–2.2) | 4 (0.65%, 0.2–1.7) | 3 (0.5%, 0.1–1.5) |
| >60 years | 161 (54.2) | 136 (45.8) | 297 (33.8) | 7 (2.4%, 1.05–4.9) | 4 (1.35%, 0.4–3.5) | 3 (1.0%, 0.2–3.1) | 154 (50.0) | 154 (50.0) | 308 (33.4) | 2 (0.65%, 0.0–2.5) | 2 (0.65%, 0.0–2.5) | 2 (0.65%, 0.0–2.5) |
| Total | 416 (47.3) | 463 (52.7) | 879 | 17 (1.9%, 1.2–3.1) | 12 (1.4%, 0.8–2.4) | 6 (0.7%, 0.3–1.5) | 461 (50.1) | 460 (49.9) | 921 | 8 (0.9%, 0.4–1.7) | 6 (0.65%, 0.3–1.45) | 5 (0.5%, 0.2–1.3) |

95% CI, 95% confidence interval; IFA, immunofluorescence assay; MN, micro neutralization; PRN, plaque reduction neutralization.

**Table 2.** Information of subjects (years of collection, age and sex) and serologic results (ELISA, IFA, MN and PRN titer) of the samples showing WNV neutralizing antibodies by MN/PRN assays.

| Years | Subject | Age Group | Age (Years) | Sex | ELISA IgG | IFA IgG | MN Titer | PRN Titer |
|---|---|---|---|---|---|---|---|---|
| 2016–2017 | 1 | >60 | 61 | F | Positive | 1/100 | 20 | 10 |
| | 2 | >60 | 64 | M | Positive | 1/6400 | 20 | 40 |
| | 3 | 20–60 | 57 | F | Borderline | 1/50 | 20 | 10 |
| | 4 | >60 | 91 | F | Borderline | 1/50 | 20 | 20 |
| | 5 | 20–60 | 42 | F | Positive | 1/50 | 10 | 10 |
| | 6 | 20–60 | 33 | F | Positive | 1/800 | 10 | 40 |
| 2018–2019 | 7 | 20–60 | 52 | M | Positive | 1/400 | 10 | 40 |
| | 8 | >60 | 75 | M | Positive | 1/50 | 10 | 20 |
| | 9 | 20–60 | 30 | M | Positive | 1/3200 | 20 | 10 |
| | 10 | 20–60 | 32 | M | Positive | 1/6400 | 40 | 40 |
| | 11 | >60 | 85 | F | Positive | 1/100 | 20 | 20 |

IFA, immunofluorescence assay; MN, micro neutralization; PRN, plaque reduction neutralization.

**Table 3.** Results for the univariate logistic regression model and the multivariate logistic regression model by independent variables (sex and age group).

| Years of Collection | Independent Variable | Univariate Logistic Regression Model | | | Multivariate Logistic Regression Model | | |
|---|---|---|---|---|---|---|---|
| | | OR | 95% CI | *p* Value | OR | 95% CI | *p* Value |
| 2016–2017 | Sex | 0.80 | 0.30–2.08 | 0.64 | 0.82 | 0.31–2.16 | 0.69 |
| | Age group | 1.39 | 0.52–3.68 | 0.51 | 1.36 | 0.51–3.62 | 0.54 |
| 2018–2019 | Sex | 0.28 | 0.06–1.37 | 0.12 | 0.28 | 0.06–1.37 | 0.12 |
| | Age group | 0.57 | 0.12–2.74 | 0.48 | 0.57 | 0.12–2.74 | 0.48 |

OR, odd ratio; 95% CI, 95% confidence interval.

## 3. Discussion

To our knowledge, this is the first study on the prevalence of WNV antibodies in the Tuscany region. The results show that, although between 2016 and 2019 WNV prevalence in the area of Siena Province was as low as less than 1%, in 2016 and 2017 WNV was actively circulating, as shown by the finding of specific IgM suggestive of recent infection. Moreover, in some of the positive subjects the presence of antibodies with neutralizing may suggest a potentially protective immunity.

To date, the presence of WNV in Tuscany has been reported in horses in 1998 [4], in 2009 [10], and in 2016 [11], while WNV infection in humans was retrospectively diagnosed in 2007 [6] and two cases of WNND were reported in 2017 [13].

This is the first study showing that cases of human infection by WNV have occurred in the Siena area and may be considered of some relevance as no human cases have been reported by routine surveillance in the study period 2016–2019. In addition, it appears that WNV circulation was not an occasional finding but was detected in both 2016–2017 and 2018–2019 periods studied. The reason why no cases of WNV infection were reported in the study period from the Siena area may be due to the fact that WNV is often the cause of mild or sub-clinical infection, which may lead to misdiagnosis and underreporting of the disease.

Other serological studies conducted on the general population or blood donors in other areas of Italy where the circulation of WNV has been registered (Lombardy, Emilia-Romagna, and Veneto regions) [20–24] have found similar prevalence to this study. The same median age of 57 years we found in our study was observed in WNV positive blood donors in the Veneto region [25], probably due to the fact that our study most likely includes asymptomatic WNV infections or mild symptomatic infections.

Similar WNV prevalence studies performed in other European countries and in the Mediterranean Basin detected neutralizing antibodies in 1.5% and 2.34% of the population in Greece [26] and in Hungary [27], respectively, while the results of this study are more in line with the prevalence observed in Bulgaria [28].

Historically, surveillance activities detected WNV infection cases in the Northern East areas of Italy. However, over the years, increasing numbers of West Nile fever cases have been reported from other areas, suggesting viral circulation expanding in the general population in areas previously considered naïve [3], as shown in our study. In fact, the population included in this study, although not selected for the purpose, better represents the general population than that included in other epidemiological studies performed in blood donors or in international travelers [21,22,24,29,30].

This study has some limitation. Samples were collected for purposes different from the aim of this study; thus, no information on clinical findings such as fever or neurological signs and symptoms was available. Our study population did not include subjects younger than 20 years of age.

Recently, among mosquito-borne flaviviruses, with birds as reservoir hosts, circulating in different areas of Europe, Usutu virus has been reported to possess serological cross-reactions with WNV [31]. In our study, WNV-positive samples were not tested for Usutu virus neutralizing antibodies; therefore, a possible cross-reactivity between the two viruses cannot be totally excluded. Usutu virus circulation in the Tuscany region has been reported only in mosquito pools in 2018 and 2019 from two provinces in the Northern area of the region, and no animal or human cases were reported from routine surveillance activities [16,32]. Moreover, IgM ELISA antibodies have been detected in some samples of this study. IgM ELISA is usually considered to be more specific than IgG, with a lower cross-reactivity with other flaviviruses [33,34]. Taking into account the data on the circulation of both viruses in the area and the results obtained from all the serological assays performed in this study, the specific reaction to WNV can be reasonably assumed.

This study shows for the first time the active circulation in humans of WNV that occurred between 2016 and 2019 in the Siena area, an area considered not at high risk until 2019. Although the prevalence of WNV is limited as compared to other neurotropic arboviruses, such as Toscana virus [35], it appears to have acquired an established transmission pattern between 2016 and 2019.

In conclusion, WNV infection appears to be more widespread in the area of Siena than has been detected so far, and it is possible that some cases of infection are underdiagnosed and underreported. Taking into consideration the trend of the expansion of WNV in Central Italy, the absence of reported WNV human cases in the Siena area should not limit the application of preventive measures and epidemiological surveillance, as the low prevalence of antibodies does not prevent outbreaks of WNV disease in the future.

## 4. Materials and Methods

### 4.1. Study Population

The study was performed with samples available at the sera bank of the Molecular Epidemiology Laboratory of the University of Siena, Italy. Human serum samples are residual samples collected from a local laboratory in the province of Siena between 2016 and 2019. Samples were anonymously collected and stored in compliance with Italian ethics law. For each serum sample, information only on age, sex, place and year of sampling was available.

A total of 1800 samples were randomly selected from the sera bank: 879 for the years 2016–2017 and 921 for the years 2018–2019.

### 4.2. ELISA and Immunofluorescence Assay

All samples were tested for the presence of IgG antibodies against WNV by use of "West Nile Virus IgG" (DIA.PRO, Milano, Italy) commercial ELISA kit. Testing was performed according to manufacturer's instructions, and test results were calculated by means of a cut-off value determined with the following formula: Cut-off = optical density (OD) of the negative control + 0.250. Samples were considered positive when the ratio between the OD of the sample and that of the cut-off was >1.1, and negative when the ratio between the OD of the sample and that of the cut-off was <0.9. Samples with a ratio between 0.9 and 1.1 were considered borderline.

Out of 879 samples collected in 2016–2017, 92 were also tested for the presence of IgM antibodies by use of "West Nile Virus IgM" (DIA.PRO, Milano, Italy) commercial ELISA kit. IgM ELISA testing was performed on all the ELISA IgG positive and borderline samples and on a subset of ELISA IgG negative samples. Testing was performed according to manufacturer's instructions and results were calculated as for ELISA IgG kit.

ELISA borderline and positive samples were further tested by "Anti-West Nile virus (IgG)" (EUROIMMUN, Lübeck, Germany) IFA commercial kit, following manufacturer's instructions. Samples were tested with 2-fold dilutions from 1:50 to 1:6400. The IFA titer was defined as the highest serum dilution showing fluorescence, as reported by manufacturer's instructions.

All IgG and IgM ELISA and IgG IFA positive samples were further tested by MN and PRN assays.

### *4.3. Micro Neutralization and Plaque Reduction Neutralization Assays*

The cell substrate used was Vero E6 (African green monkey kidney cell line; ATCC® CRL-1586™) propagated in Dulbecco's Modified Eagle's Medium (DMEM; Sigma-Aldrich, St. Louis, MO, USA) supplemented with 10% Fetal Bovine Serum (FBS; Sigma-Aldrich, St. Louis, MO, USA). The WNV strain (lineage 2) viral stock, consisting of cell-free supernatants of acutely infected Vero E6 cells, was stored at −80 °C until use. Prior to the MN and PRN test, WNV was titrated for 50% tissue culture infectious dose ($TCID_{50}$) and plaque forming unit (PFU) using Vero E6 cells, and all serum samples were heat-inactivated at 56 °C for 30 min.

MN assay was performed by exposing (1:1) serial twofold dilutions of heat-inactivated serum in DMEM (1:10 to 1:320) to 100 $TCID_{50}$ of WNV. After 1-h incubation at 37 °C in 5% $CO_2$ atmosphere, 50 μL of the serum/virus mixture was plated on each well of a 96-well plate covered by Vero E6 cell monolayers ($10^4$ cells/well), and incubated for 1 h at 37 °C, 5% $CO_2$. Then, 50 μL of DMEM was added on each well and the plate was incubated for 4 days up to the appearance of an easy detectable cytopathic effect in control cultures (cell monolayers exposed to WNV). Additionally, IgG serum negative to WNV was used as control. The antibody titer was defined as the reciprocal of the highest dilution of the test serum sample, which showed at least 50% neutralization.

PRN assay was performed on heat-inactivated serum samples by exposing (1:1) serial twofold dilutions of them in DMEM (1:10 to 1:320) to 100 PFU of WNV. After incubation for 1 h at 37 °C, 5% $CO_2$ atmosphere, 300 μL of the serum/virus mixture was plated on each well of 6-well plates seeded with $2.5 \times 10^5$ Vero E6 cells and incubated 1 h at 37 °C. Then, the overlay medium composed of 0.5% Sea Plaque Agarose (Lonza, Basel, Switzerland) diluted in propagation medium was added to each well. After 4 days of incubation at 37 °C, the monolayers were fixed with methanol (Carlo Erba Chemicals, Milan, Italy) and stained with 0.1% crystal violet (Carlo Erba Chemicals, Milan, Italy) and the viral titers were calculated by PFU counting. Percent of PRN was calculated by dividing the average PFU of viral serum treated samples by the average of viral positive control. All experiments were repeated at least twice. All experimental procedures were conducted under biosafety level 3 containment.

### *4.4. Statistical Analysis*

Categorical dichotomous data (sex and age group) and discrete data (IgG ELISA, IFA, MN/PRN assays results) were defined as categorical dichotomous data, described as counts and percentages and evaluated by Chi-square test. The relations between the IgG positivity of each assay as a dependent categorical dichotomous variable defined as a dummy variable and independent factors (sex and age group) were evaluated by logistic regression model, and OR, 95% CI, and *p*-values were assessed. In the univariate logistic regression model, all the factors related to IgG positivity were investigated as independent variables. The statistically significant independent variables were assessed in the multivariate logistic regression model using Wald test and stepwise method for the selection of $p$-value. Statistical significance was set at $p < 0.05$.

Data from statistical analyses were performed with the software GraphPad Prism v.6.0.0 (GraphPad Software, San Diego, CA, USA).

### 4.5. Geographic Methods

The spatial distribution of WND human and equine reported cases was mapped using QGIS 3.6.0 [36]. The shapefile of Tuscany region (WGS84 UTM32N) was retrieved from the National Institute of Statistics (ISTAT) [37]. The national geographic map was used as basemap to relate the study area to the European region.

**Author Contributions:** Conceptualization, S.M. and S.V.; methodology, S.M., S.G., M.A.S.; formal analysis, S.M., G.L., M.C.; investigation, S.M., S.G., M.A.S., C.A., R.C.; writing—original draft preparation, S.M.; writing—review and editing, E.M., S.V., S.G., M.A.S., G.L., M.C., C.M.T. All authors have read and agreed to the published version of the manuscript.

**Funding:** This research received no external funding.

**Institutional Review Board Statement:** Not applicable.

**Informed Consent Statement:** Not applicable.

**Conflicts of Interest:** The authors declare no conflict of interest.

## References

1. Beck, C.; Jimenez-Clavero, M.A.; Leblond, A.; Durand, B.; Nowotny, N.; Leparc-Goffart, I.; Zientara, S.; Jourdain, E.; Lecollinet, S. Flaviviruses in Europe: Complex Circulation Patterns and Their Consequences for the Diagnosis and Control of West Nile Disease. *Int. J. Environ. Res. Public Health* **2013**, *10*, 6049–6083. [CrossRef]
2. Gould, E.; Solomon, T. Pathogenic flaviviruses. *Lancet* **2008**, *371*, 500–509. [CrossRef]
3. Rizzo, C.; Napoli, C.; Venturi, G.; Pupella, S.; Lombardini, L.; Calistri, P.; Monaco, F.; Cagarelli, R.; Angelini, P.; Bellini, R.; et al. West Nile virus transmission: Results from the integrated surveillance system in Italy, 2008 to 2015. *Eurosurveillance* **2016**, *21*, 30340. [CrossRef]
4. Autorino, G.L.; Battisti, A.; Deubel, V.; Ferrari, G.; Forletta, R.; Giovannini, A.; Lelli, R.; Murri, S.; Scicluna, M.T. West Nile virus Epidemic in Horses, Tuscany Region, Italy. *Emerg. Infect. Dis.* **2002**, *8*, 1372–1378. [CrossRef]
5. Rossini, G.; Cavrini, F.; Pierro, A.; Macini, P.; Finarelli, A.C.; Po, C.; Peroni, G.; Di Caro, A.; Capobianchi, M.R.; Nicoletti, L.; et al. First human case of West Nile virus neuroinvasive infection in Italy, September 2008—Case report. *Eurosurveillance* **2008**, *13*, 19002. [CrossRef]
6. Cusi, M.G.; Roggi, A.; Terrosi, C.; Savellini, G.G.; Toti, M. Retrospective Diagnosis of West Nile Virus Infection in a Patient with Meningoencephalitis in Tuscany, Italy. *Vector Borne Zoonotic Dis.* **2011**, *11*, 1511–1512. [CrossRef] [PubMed]
7. Rizzo, C.; Vescio, F.; Declich, S.; Finarelli, A.C.; Macini, P.; Mattivi, A.; Rossini, G.; Piovesan, C.; Barzon, L.; Palù, G.; et al. West Nile virus transmission with human cases in Italy, August–September 2009. *Eurosurveillance* **2009**, *14*, 19353. [CrossRef]
8. Ministero della Salute. Sorveglianza della Malattia di West Nile in Italia—2010. Available online: https://www.trovanorme.salute.gov.it/norme/renderNormsanPdf?anno=0&codLeg=34923&parte=1%20&serie= (accessed on 15 March 2021).
9. Moirano, G.; Richiardi, L.; Calzolari, M.; Merletti, F.; Maule, M. Recent rapid changes in the spatio-temporal distribution of West Nile Neuro-invasive Disease in Italy. *Zoonoses Public Health* **2019**, *67*, 54–61. [CrossRef]
10. Monaco, F.; Savini, G.; Calistri, P.; Polci, A.; Pinoni, C.; Bruno, R.; Lelli, R. 2009 West Nile disease epidemic in Italy: First evidence of overwintering in Western Europe? *Res. Vet. Sci.* **2011**, *91*, 321–326. [CrossRef] [PubMed]
11. Scaramozzino, P.; Carvelli, A.; Bruni, G.; Cappiello, G.; Censi, F.; Magliano, A.; Manna, G.; Ricci, I.; Rombolà, P.; Romiti, F.; et al. West Nile and Usutu viruses co-circulation in central Italy: Outcomes of the 2018 integrated surveillance. *Parasites Vectors* **2021**, *14*, 243. [CrossRef]
12. Rizzo, C.; Salcuni, P.; Nicoletti, L.; Ciufolini, M.G.; Russo, F.; Masala, R.; Frongia, O.; Finarelli, A.C.; Gramegna, M.; Gallo, L.; et al. Epidemiological surveillance of West Nile neuroinvasive diseases in Italy, 2008 to 2011. *Eurosurveillance* **2012**, *17*, 20172. [CrossRef] [PubMed]
13. Istituto Superiore di Sanità. Sorveglianza Integrata del West Nile e Usutu Virus, Bollettino N. 13 del 9 Novembre 2017. Available online: https://www.epicentro.iss.it/westnile/bollettino/Bollettino%20WND_08.11.2017.pdf (accessed on 15 March 2021).
14. Istituto Superiore di Sanità. Sorveglianza integrata del West Nile e Usutu Virus, Bollettino N. 1 del 28 Giugno 2018. Available online: https://www.epicentro.iss.it/westnile/bollettino/Bollettino%20WND_n.1%2028.06.2018.pdf (accessed on 15 March 2021).

15. Riccardo, F.; Monaco, F.; Bella, A.; Savini, G.; Russo, F.; Cagarelli, R.; Dottori, M.; Rizzo, C.; Venturi, G.; Di Luca, M.; et al. An early start of West Nile virus seasonal transmission: The added value of One Heath surveillance in detecting early circulation and triggering timely response in Italy, June to July 2018. *Eurosurveillance* **2018**, *23*, 1800427. [CrossRef]
16. Istituto Superiore di Sanità. Sorveglianza Integrata del West Nile e Usutu virus, Bollettino N. 18 del 15 Novembre 2018. Available online: https://www.epicentro.iss.it/westnile/bollettino/Bollettino%20WND_%20N.%2018%20%2015.%2011%202018.pdf (accessed on 15 March 2021).
17. Ministero della Salute. Piano Nazionale Integrato di Sorveglianza e Risposta ai Virus West Nile e Usutu—2018. Available online: https://www.trovanorme.salute.gov.it/norme/renderNormsanPdf?anno=2018&codLeg=65084&parte=1%20&serie=null (accessed on 15 March 2021).
18. Ministero della Salute. Piano Nazionale Integrato di Prevenzione, Sorveglianza e Risposta ai Virus West Nile e Usutu—2019. Available online: https://www.trovanorme.salute.gov.it/norme/renderNormsanPdf?anno=2019&codLeg=68806&parte=1%20&serie=null (accessed on 15 March 2021).
19. Ministero della Salute. Piano Nazionale di Prevenzione, Sorveglianza e Risposta alle Arbovirosi (PNA) 2020–2025. Available online: https://www.salute.gov.it/imgs/C_17_pubblicazioni_2947_allegato.pdf (accessed on 15 March 2021).
20. Faggioni, G.; De Santis, R.; Pomponi, A.; Grottola, A.; Serpini, G.F.; Meacci, M.; Gennari, W.; Tagliazucchi, S.; Pecorari, M.; Monaco, F.; et al. Prevalence of Usutu and West Nile virus antibodies in human sera, Modena, Italy, 2012. *J. Med. Virol.* **2018**, *90*, 1666–1668. [CrossRef]
21. Pierro, A.; Gaibani, P.; Manisera, C.; Dirani, G.; Rossini, G.; Cavrini, F.; Ghinelli, F.; Ghinelli, P.; Finarelli, A.C.; Mattivi, A.; et al. Seroprevalence of West Nile Virus-Specific Antibodies in a Cohort of Blood Donors in Northeastern Italy. *Vector Borne Zoonotic Dis.* **2011**, *11*, 1605–1607. [CrossRef]
22. Pierro, A.; Gaibani, P.; Spadafora, C.; Ruggeri, D.; Randi, V.; Parenti, S.; Finarelli, A.C.; Rossini, G.; Landini, M.P.; Sambri, V. Detection of specific antibodies against West Nile and Usutu viruses in healthy blood donors in northern Italy, 2010–2011. *Clin. Microbiol. Infect.* **2013**, *19*, E451–E453. [CrossRef]
23. Gaibani, P.; Pierro, A.; Lunghi, G.; Farina, C.; Toschi, V.; Matinato, C.; Orlandi, A.; Zoccoli, A.; Almini, D.; Landini, M.P.; et al. Seroprevalence of West Nile virus antibodies in blood donors living in the metropolitan area of Milan, Italy, 2009–2011. *New Microbiol.* **2013**, *36*, 81–83.
24. Pezzotti, P.; Piovesan, C.; Barzon, L.; Cusinato, R.; Cattai, M.; Pacenti, M.; Piazza, A.; Franchin, E.; Pagni, S.; Bressan, S.; et al. Prevalence of IgM and IgG antibodies to West Nile virus among blood donors in an affected area of north-eastern Italy, summer 2009. *Eurosurveillance* **2011**, *16*, 19814. [CrossRef]
25. Barzon, L.; Pacenti, M.; Franchin, E.; Pagni, S.; Lavezzo, E.; Squarzon, L.; Martello, T.; Russo, F.; Nicoletti, L.; Rezza, G.; et al. Large Human Outbreak of West Nile Virus Infection in North-Eastern Italy in 2012. *Viruses* **2013**, *5*, 2825–2839. [CrossRef]
26. Hadjichristodoulou, C.; Pournaras, S.; Mavrouli, M.; Marka, A.; Tserkezou, P.; Baka, A.; Billinis, C.; Katsioulis, A.; Psaroulaki, A.; Papa, A.; et al. West Nile Virus Seroprevalence in the Greek Population in 2013: A Nationwide Cross-Sectional Survey. *PLoS ONE* **2015**, *10*, e0143803. [CrossRef]
27. Nagy, A.; Szöllősi, T.; Takács, M.; Magyar, N.; Barabás, É. West Nile Virus Seroprevalence Among Blood Donors in Hungary. *Vector Borne Zoonotic Dis.* **2019**, *19*, 844–850. [CrossRef]
28. Christova, I.; Panayotova, E.; Tchakarova, S.; Taseva, E.; Trifonova, I.; Gladnishka, T. A nationwide seroprevalence screening for West Nile virus and Tick-borne encephalitis virus in the population of Bulgaria. *J. Med. Virol.* **2017**, *89*, 1875–1878. [CrossRef] [PubMed]
29. Loconsole, D.; Metallo, A.; De Robertis, A.L.; Morea, A.; Quarto, M.; Chironna, M. Seroprevalence of Dengue Virus, West Nile Virus, Chikungunya Virus, and Zika Virus in International Travelers Attending a Travel and Migration Center in 2015–2017, Southern Italy. *Vector Borne Zoonotic Dis.* **2018**, *18*, 331–334. [CrossRef] [PubMed]
30. Barzon, L.; Pacenti, M.; Cusinato, R.; Cattai, M.; Franchin, E.; Pagni, S.; Martello, T.; Bressan, S.; Squarzon, L.; Cattelan, A.M.; et al. Human cases of West Nile Virus Infection in north-eastern Italy, 15 June to 15 November 2010. *Eurosurveillance* **2011**, *16*, 19949. [CrossRef]
31. Llorente, F.; García-Irazábal, A.; Pérez-Ramírez, E.; Cano-Gómez, C.; Sarasa, M.; Vázquez, A.; Jiménez-Clavero, M.Á. Influence of flavivirus co-circulation in serological diagnostics and surveillance: A model of study using West Nile, Usutu and Bagaza viruses. *Transbound. Emerg. Dis.* **2019**, *66*, 2100–2106. [CrossRef] [PubMed]
32. Istituto Superiore di Sanità. Sorveglianza Integrata del West Nile e Usutu Virus, Bollettino N. 16 del 25 Novembre 2019. Available online: https://www.epicentro.iss.it/westnile/bollettino/Bollettino-WND-N16-25nov2019.pdf (accessed on 1 June 2021).
33. Tardei, G.; Ruta, S.; Chitu, V.; Rossi, C.; Tsai, T.F.; Cernescu, C. Evaluation of immunoglobulin M (IgM) and IgG enzyme immunoassays in serologic diagnosis of West Nile Virus infection. *J. Clin. Microbiol.* **2000**, *38*, 2232–2239. [CrossRef]
34. Martin, D.A.; Biggerstaff, B.J.; Allen, B.; Johnson, A.J.; Lanciotti, R.S.; Roehrig, J.T. Use of immunoglobulin m cross-reactions in differential diagnosis of human flaviviral encephalitis infections in the United States. *Clin. Diagn. Lab. Immunol.* **2002**, *9*, 544–549. [CrossRef]
35. Marchi, S.; Trombetta, C.M.; Kistner, O.; Montomoli, E. Seroprevalence study of Toscana virus and viruses belonging to the Sandfly fever Naples antigenic complex in central and southern Italy. *J. Infect. Public Health* **2017**, *10*, 866–869. [CrossRef]

36. QGIS Development Team. Quantum GIS Geographic Information System. Open Source Geospatial Foundation Project. 2017. Available online: http://qgis.osgeo.org. (accessed on 8 June 2021).
37. ISTAT. Confini delle Unità Amministrative a Fini Statistici al 1° Gennaio 2021. Available online: https://www.istat.it/it/archivio/222527 (accessed on 8 June 2021).

*pathogens*

MDPI

*Article*

# First Autochthonous West Nile Lineage 2 and Usutu Virus Infections in Humans, July to October 2018, Czech Republic

**Hana Zelená [1,2], Jana Kleinerová [3], Silvie Šikutová [4], Petra Straková [4,5], Hana Kocourková [6,7], Roman Stebel [6,7], Petr Husa [6,7], Petr Husa, Jr. [6,7], Eva Tesařová [8], Hana Lejdarová [9], Oldřich Šebesta [10], Peter Juráš [10], Renata Ciupek [10], Jakub Mrázek [1] and Ivo Rudolf [4,*]**

1 Public Health Institute, Partyzánské nám. 7, 702 00 Ostrava, Czech Republic; hana.zelena@zuova.cz (H.Z.); jakub.mrazek@zuova.cz (J.M.)
2 Department of Biomedical Sciences, Faculty of Medicine, University of Ostrava, Syllabova 19, 703 00 Ostrava, Czech Republic
3 Department of Infectious Diseases, Hospital Břeclav, U Nemocnice 3066/1, 690 74 Břeclav, Czech Republic; jana.kleinerova@post.cz
4 Institute of Vertebrate Biology, The Czech Academy of Sciences, Květná 8, 603 65 Brno, Czech Republic; sikutova@ivb.cz (S.Š.); strakova.p@centrum.cz (P.S.)
5 Veterinary Research Institute, Hudcova 70, 621 00 Brno, Czech Republic
6 Department of Infectious Diseases, University Hospital Brno, Jihlavská 20, 602 00 Brno, Czech Republic; kocourkova.hana@fnbrno.cz (H.K.); stebel.roman@fnbrno.cz (R.S.); husa.petr@fnbrno.cz (P.H.); husa.petr2@fnbrno.cz (P.H.J.)
7 Department of Infectious Diseases, Faculty of Medicine, Masaryk University, Kamenice 753/5, 625 00 Brno, Czech Republic
8 Department of Health Insurance, University Hospital Brno, Jihlavská 20, 602 00 Brno, Czech Republic; tesarova.eva@fnbrno.cz
9 Transfusion and Tissue Department, University Hospital Brno, Jihlavská 20, 602 00 Brno, Czech Republic; lejdarova.hana@fnbrno.cz
10 Regional Public Health Authority of the Southern Moravia Region, Jeřábkova 4, 602 00 Brno, Czech Republic; oldrich.sebesta@tiscali.cz (O.Š.); peter.juras@khsbrno.cz (P.J.); renata.ciupek@khsbrno.cz (R.C.)
* Correspondence: rudolf@ivb.cz; Tel.: +420-519-352-961; Fax: +420-519-352-387

**Citation:** Zelená, H.; Kleinerová, J.; Šikutová, S.; Straková, P.; Kocourková, H.; Stebel, R.; Husa, P.; Husa, P., Jr.; Tesařová, E.; Lejdarová, H.; et al. First Autochthonous West Nile Lineage 2 and Usutu Virus Infections in Humans, July to October 2018, Czech Republic. *Pathogens* **2021**, *10*, 651. https://doi.org/10.3390/pathogens10060651

Academic Editor: Francisco Llorente

Received: 28 April 2021
Accepted: 21 May 2021
Published: 24 May 2021



**Abstract:** We present epidemiological, clinical and laboratory findings of five Czech patients diagnosed with autochthonous mosquito-borne disease—four patients with confirmed West Nile virus (WNV) and one patient with Usutu virus (USUV) infections, from July to October 2018, including one fatal case due to WNV. This is the first documented human outbreak caused by WNV lineage 2 in the Czech Republic and the first record of a neuroinvasive human disease caused by USUV, which illustrates the simultaneous circulation of WNV and USUV in the country.

**Keywords:** West Nile virus; Usutu virus; mosquito-borne infections; human

## 1. Introduction

West Nile virus (WNV) is a mosquito-borne arbovirus belonging to the family *Flaviviridae*, genus *Flavivirus*, Japanese encephalitis serocomplex. In nature, it circulates mainly among birds and mosquitoes of genera *Culex*, *Anopheles*, *Culiseta*, *Uranotaenia* or *Coquilletidia*. WNV is a causative agent of West Nile fever, a mosquito-borne disease affecting horses and humans, the latter serving as so-called dead-end hosts [1]. Since 2004, highly virulent WNV lineage 2 has appeared in Europe, causing sporadic outbreaks in Hungary (2008), Greece (2010) and Serbia (2012) [2]. In the Czech Republic, WNV research in birds, mosquito vectors and humans has a long tradition. Importantly, highly virulent WNV lineage 2 (WNV-2) strains have been repeatedly documented in *Cx. modestus* populations on local fishponds [3,4]. Interestingly, no case of WNV-2 infection in humans had been documented before the 2018 season.

Usutu virus is a mosquito-borne flavivirus (family *Flaviviridae*, genus *Flavivirus*), in the Japanese encephalitis virus serocomplex. Like WNV, it circulates among birds and ornithophilic mosquitoes, with demonstrated pathogenicity to a wide variety of wild and domestic birds. It emerged in 2001 in Austria with the highest mortality recorded in blackbirds [5]. Since then USUV has spread further to central and western European countries including the Czech Republic, where it has established itself among blackbirds [6,7] and mosquitoes [8]. Since its introduction in Europe, there have been several reports of neuroinvasive disease in immunocompromised [9] and immunocompetent humans [10,11]. Similar to WNV, USUV has been detected in asymptomatic blood donors in Italy [12], Austria [13] and Germany [14].

The aim of this report was to summarize the epidemiological and clinical characteristics of patients diagnosed by infection with either WNV-2 or USUV in the Czech Republic and to expand our knowledge of these emerging mosquito-borne diseases in Central Europe.

## 2. Case Reports

We summarized epidemiological, clinical and laboratory findings collected from five patients diagnosed with WNV or USUV infections. Specific data are summarized in Tables 1–3. All but one had no history of traveling to a known WNV endemic area. One patient (Case 1) reported previous vaccination against tick-borne encephalitis virus (TBEV), a related tick-borne flavivirus circulating in the area, that might present with cross-reactive immunity, which could subsequently complicate an accurate final diagnostic result. All patients reside in South Moravia.

**Table 1.** Descriptive epidemiological and clinical summary of five patients diagnosed with mosquito-borne disease, from July to October 2018, Czech Republic.

| | Case 1 | Case 2 | Case 3 | Case 4 | Case 5 |
|---|---|---|---|---|---|
| **Demographic data** | | | | | |
| Age (years) | 74 | 72 | 51 | 52 | 46 |
| Gender | Male | Female | Male | Male | Female |
| **Epidemiological data** | | | | | |
| Area of residence | Urban | Urban | Urban | Urban | Urban |
| Occupation | Retired | Retired | Construction Manager | Sales Representative | Shop Assistant |
| Travel history | Austria (1 day) | No | No | No | Turkey |
| Outdoor activity | fishing | gardening | walking | gardening | gardening |
| Contact with mosquitoes | Yes | Yes | Yes | Yes | Yes |
| Vaccination status (TBE, YF) | Yes (TBE) | No | No | No | No |
| Blood donation/Transfusion | No | No | No | No | No |
| **Clinical presentation** | | | | | |
| Date of disease onset | 23 July 2018 | 7 August 2018 | 10 September 2018 | 11 September 2018 | 22 September 2018 |
| Days of hospitalization | N/A | 15 | 14 | 17 | 12 |
| Main clinical signs and symptoms | Fever, headache, muscle pain, diarrhea, macular exanthema | Fever, diarrhea, muscle pain, weakness | Fever, headache, arthralgia, fatigue | Fever, fatigue, diarrhea, arthralgia, headache, vomiting | Fever, headache, vomiting, ataxia, meningeal signs |
| The highest body temperature | 40 °C | 39 °C (anamnestic) | 37.6 °C | 39.4 °C | Not measured at home |
| The lowest GCS | 15 | 3 | 15 | 15 | 14 |
| Comorbidities | Ischemic heart disease, arterial hypertension, HLA-B27 positive | Rheumatoid arthritis, Lichen ruber, chronic gastritis, hypothyroidism | Vertebrogenic algic syndrome | Chronic cefalea, hypertension | Gastroesophageal reflux disease |
| Clinical diagnosis | Fever | Meningoencephalitis | Meningitis | Meningitis | Meningitis |
| Duration of disease (No. days) | 13 | 20 | 36 | 24 | 22 |
| Outcome at discharge (GOS) | Recovered (8) | Deceased | Recovered (8) | Recovered (8) | Recovered (8) |

Legend: TBE-Tick-borne encephalitis; YF-Yellow fever; GCS-Glasgow Coma Scale; GOS-Glasgow Outcome Scale; NA—not applicable.

**Table 2.** Laboratory and neuropathological findings of five patients diagnosed with mosquito-borne disease, from July to October 2018, Czech Republic.

| | Case 1 | Case 2 | Case 3 | Case 4 | Case 5 | Reference Range |
|---|---|---|---|---|---|---|
| **Cerebrospinal fluid (CSF) examination** | Day 9 | Day 13 | Day 22 | Day 8 | Day 10 | |
| Cell count/$mm^3$ | - | 5 | 12 | 15 | 115 | 0–5 |
| Polymorphonuclear/mononuclear cells | - | - | - | - | 5/110 | - |
| Proteins (g/L) | - | 0.6 | - | 0.99 | 0.87 | 0.15–0.45 |
| Glucose (mmol/L) | - | 4.3 | - | 4.3 | 2.9 | N/A |
| Lactate (mmol/L) | - | - | - | 2.9 | 2.7 | 1.1–2.4 |
| **Serum examination** | | | | | | |
| C-reactive protein; CRP (mg/L) | 1.4 | 25 | 12 | 5 | 22.7 | 0-5 |
| White blood cells; WBC ($\times 10^9$/L) | 8.54 | 16.22 | 7.19 | 8.05 | 5.19 | 4–10 |
| Platelets ($\times 10^9$/L) | 243 | 310 | 225 | 118 | 244 | 150–400 |
| Red blood cells; RBC ($\times 10^{12}$/L) | 4.88 | 5.24 | 5.29 | 4.44 | 4.08 | 4.0–5.8 |
| Hemoglobin (g/L) | 152 | 154 | 150 | 138 | 125 | 135–175 |
| Bilirubin (μmol/L) | 11.5 | 4.9 | - | 21.0 | 7.6 | 2–21 |
| Aspartate-aminotransferase; AST (μkat/L) | 0.48 | 0.77 | - | 0.82 | 0.35 | 0.17–0.85 |
| Alanine-aminotransferase; ALT (μkat/L) | 0.51 | 0.81 | - | 2.25 | 0.28 | 0.17–0.83 |
| Gamma-glutamyltransferase; GGT (μkat/L) | 1.20 | - | - | - | 0.69 | 0.13–1.02 |
| Lactate dehydrogenase; LD (μkat/L) | 4.08 | - | - | - | N/A | 2.25–3.75 |
| **Brain examination/imaging** | | | | | | |
| Brain computed tomography; CT | - | hypodensity vs.postinfectious | no pathological findings | onset of oedema | - | - |
| Brain magnetic resonance imaging; MRI | - | - | - | without signs of inflammation | - | - |

**Table 3.** Diagnostic outcomes of five patients diagnosed with mosquito-borne disease, from July to October 2018, Czech Republic.

| Case | Days Tested | WNV PCR (S,WB, CSF,U) | IgM [1] WNV ELISA Euroimmun [2] WNV ELISA Focus [3] WNV IIFT Euroimmun | IgG [1] WNV ELISA Euroimmun [2] WNV ELISA Focus [3] WNV IIFT Euroimmun | WNV VNT (titre) | [4] TBEV IgM ELISA (IP) | [4] TBEV IgG ELISA (IP) | [4] TBEV IgG avidity (%) | TBEV VNT (Titre) | USUV VNT (Titre) | Case Status |
|---|---|---|---|---|---|---|---|---|---|---|---|
| 1 | 23.10. 2018 | | 0.91 (eq.) [1]<br>**1.57 (pos.)** [2] | **3.49 (pos)** [1] | **512 (pos.)** | neg. | **12.67 (pos.)** | 100% | 32 (pos.) | **128 (pos.)** | WNV confirmed |
| 2 | 18.8. 2018 | | **>20 (pos.)** [3] | **>20 (pos.)** [3] | 4 (eq.) | neg. | neg. | | neg. | | WNV confirmed |
| | 20.8. 2018 | S-inhib.<br>**CSF-pos.** | **>20 (pos.)** [3] | **>20 (pos.)** [3] | **8 (pos.)** | | | | | | |
| | 24.8. 2018 | **S-pos.**<br>CSF-neg. | **>20 (pos.)** [3] | **>20 (pos.)** [3] | **16 (pos.)** | | | | | | |
| | 28.8. 2018 | **U-pos.** | | | | | | | | | |
| 3 | 3.10. 2018 | S,CSF,U-neg. | **3.11 (pos.)** [1]<br>**7.22 (pos.)** [2]<br>**>20 (pos.)** [3] | 1.10 (eq.) [1]<br>1.02 (eq.) [2]<br>**>20 (pos.)** [3] | 64 (pos.) | neg. | neg. | | neg. | **16 (pos.)** | WNV confirmed |
| | 4.10. 2018 | **WB-pos.** | | | | | | | | | |
| | 10.10. 2018 | **WB-pos.**<br>S,U-neg. | **5.63 (pos.)** [2] | **2.28 (pos.)** [2] | **128 (pos.)** | neg. | neg. | | neg. | **32 (pos.)** | |
| | 8.11. 2018 | **WB-pos.**<br>S,U-neg. | **1.65 (pos.)** [1] | **2.82 (pos.)** [1] | **32 (pos.)** | neg. | **1.81 (pos.)** | 33 % | neg. | **8 (pos.)** | |
| | 13.12. 2018 | **WB-pos.**<br>U-neg. | **1.61(pos.)** [1] | **3.13 (pos.)** [1] | **32 (pos.)** | neg. | **2.41 (pos.)** | 36 % | neg. | neg. | |
| | 15.2. 2019 | **WB-pos.**;U-neg. | | | | | | | | | |

**Table 3.** *Cont.*

| Case | Days Tested | WNV PCR (S,WB, CSF,U) | IgM [1] WNV ELISA Euroimmun [2] WNV ELISA Focus [3] WNV IIFT Euroimmun | IgG [1] WNV ELISA Euroimmun [2] WNV ELISA Focus [3] WNV IIFT Euroimmun | WNV VNT (titre) | [4] TBEV IgM ELISA (IP) | [4] TBEV IgG ELISA (IP) | [4] TBEV IgG avidity (%) | TBEV VNT (Titre) | USUV VNT (Titre) | Case Status |
|---|---|---|---|---|---|---|---|---|---|---|---|
| 4 | 18.9. 2018 | S-neg. | **2.36 (pos.)** [1]<br>**5.27 (pos.)** [2]<br>**>500 (pos.)** [3] | neg. [1]<br>neg. [2]<br>**20 (pos.)** [3] | **neg.** | neg. | neg. | | neg. | neg. | WNV confirmed |
| | 24.9. 2018 | S-neg.<br>**U-pos.** | **2.90 (pos.)** [1]<br>**6.39 (pos.)** [2]<br>**500 (pos.)** [3] | **1.31 (pos.)** [1]<br>**1.26 (pos.)** [2]<br>**>20 (pos.)** [3] | **8 (pos.)** | 0.96 (eq.) | **1.19 (pos.)** | 38 % | neg. | 4 (eq.) | |
| | 1.10. 2018 | S, U-neg. | **3.69 (pos.)** [1]<br>**6.28 (pos.)** [2] | **2.12 (pos.)** [1]<br>**2.13 (pos.)** [2] | **64 (pos.)** | neg. | **2.18 (pos.)** | 22 % | neg. | **16 (pos.)** | |
| | 4.10. 2018 | **WB-pos.** | | | | | | | | | |
| | 19.10. 2018 | WB, U-neg. | **3.10 (pos.)** [1] | **3.16 (pos.)** [1] | **64 (pos.)** | neg. | **2.75 (pos.)** | 20 % | neg. | 4 (eq.) | |
| | 25.1. 2019 | WB neg.<br>**U-pos.** | neg. [1] | **3.05 (pos.)** [1] | **32 (pos.)** | neg. | **2.10 (pos.)** | 43 % | neg. | neg. | |
| 5 | 1.10. 2018 | S, CSF-neg. | neg. [1,2,3] | neg. [1,2,3] | neg. | neg. | neg. | | neg. | **16 (pos.)** | USUV |
| | 5.10. 2018 | WB, U-neg. | neg. [1,2]<br>**>20 (pos.)** [3] | neg. [1,2,3] | | | | | | | |
| | 10.10. 2018 | | **1.37 (pos.)** [1]<br>**20 (pos.)** [3] | neg. [1]<br>**20 (pos.)** [3] | **16 (pos.)** | neg. | neg. | | neg. | **256 (pos.)** | |
| | 25.10. 2018 | | neg. [2] | 0.93 (eq.) [1] | | | | | | | |

Legend: S-serum, WB-whole blood, CSF-cerebrospinal fluid, U-urine, IP-positivity index, IIFT-indirect immunofluorescence, VNT-virus neutralization test, pos.-positive, neg.-negative, eq.-equivocal, inhib.-inhibition; Reference values for WNV ELISA Euroimmun IgG and IgM (IP): <0.80 negative, 0.80–1.10 equivocal, >1.10 positive; Reference values for WNV ELISA Focus IgG (IP) <1.30 negative, 1.30–1.50 equivocal, >1.50 positive; Reference values for WNV ELISA Focus IgM (IP) <0.90 negative, 0.90–1.10 equivocal, >1.10 positive; Reference values for TBEV ELISA IgG and IgM (IP): <0.90 negative, 0.90–1.10 equivocal, >1.10 positive; Reference values for TBEV ELISA IgG avidity (%): <40 % low, 40–60 equivocal, >60 high; Reference values for WNV IIFT Euroimmun: <20 negative, ≥20 positive; Reference values for VNT( titre): <4 negative, 4 equivocal, >4 positive. All positive values are highlighted in bold. [1] West Nile Virus ELISA IgG, IgM ELISA, Euroimmun, Lübeck, Germany (Cat Nr. EI 2662-9601 G,M); [2] West Nile Virus ELISA IgG, IgM capture Dx SelectTM ELISA, Focus Diagnostics, Cypress, California, U.S. (Cat Nr EL 0300 G,M); [3] West Nile Virus IgG, IgM IIFT, Euroimmun, Lübeck, Germany (Cat Nr. FI 2662-1005 G,M); [4] ELISA-Viditest anti-TBEV IgG, IgG avidity and IgM, Vidia, Vestec, Czech Republic (Cat Nr. OD-170, OD-194).

### 2.1. Description of Cases

#### 2.1.1. Case 1

On 3 August 2018, a 74-year-old man was examined at the Infectious Diseases Clinic of Brno University Hospital for fever up to 40 °C for 12 days. Fever was accompanied by flu-like symptoms and severe headaches. The patient reported diarrhea during the first 4 days and observed macular rash on the skin on the 6th day after the onset of the fever. Apart from dehydration with borderline hypotension, the clinical examination revealed no obvious pathology. The patient had no symptoms of meningitis; no cerebrospinal fluid was tested. Basic laboratory samples were taken as early as 31 July 2018 (see Table 2). No increase in inflammatory parameters, no leukocytosis and no other pathology was found. Treatment was symptomatic with antipyretics, analgesics and hydration by infusion; there was no antibiotic therapy. A clinical follow-up was performed after 4 days, rash and fever subsided and temperatures also decreased after 13 days. Exhaustion and occasional headache persisted. Further stool and urine cultures were negative, as were CMV serology and stool virology. Inflammatory parameters remained low. In order to exclude infectious foci, a number of additional examinations were performed (chest X-ray, abdominal and intestinal ultrasound, dentistry, echocardiography). The only significant finding was tooth decay; the dentist recommended the extraction of several teeth over time, but ruled out the odontogenic etiology of the fever. The patient's exhaustion and fatigue slowly disappeared, the fever did not return, and recovery lasted approximately 7 weeks. The man was not hospitalized; he was in the home care of his wife—a doctor. The patient gradually returned to regular activities, including recreational sport. The diagnosis of West Nile Fever was made ex post. The serological finding of WNV correlated with the protracted clinical course of the febrile viral disease.

**Epidemiological background:** The patient's place of residence is metropolitan Brno. From June 2018 the patient slept outside on the covered terrace of a family villa. The patient stayed in Brno—except for a day trip on 6 May 2018 to the aqua park in Laa an der Thaya (Austria). From 31 May 2018 to 20 July 2018 the patient spent a total of 10 days in a cottage by a pond north of Brno. The patient was vaccinated against TBEV.

#### 2.1.2. Case 2

The second case report concerns a 72-year-old female patient admitted on 13 August 2018 at the internal department of the Břeclav Hospital for general weakness, diarrhea and myalgia lasting 6 days. She was a polymorbid patient treated for rheumatoid arthritis, lichen planus, chronic gastritis and thyroiditis. On the day of admission, the patient was found at home and could not move. She had the status of odontogenic etiology of fever in the internal medicine department. The patient gradually deteriorated, and a quantitative impairment of consciousness appeared on the 6th day of hospitalization (18 August 2018). The patient was transported to the department of anesthesiology, intensive care medicine and resuscitation where she was promptly intubated, artificial sleep was induced and artificial lung ventilation was started. Computed tomography (CT) of the brain showed non-specific hypodense areas in brain tissue. Furthermore, lumbar puncture was supplemented as part of differential diagnosis of the disturbance of consciousness. Biochemical and cytological findings in cerebrospinal fluid corresponded to aseptic neuroinfection. Serology showed borderline IgM and weakly positive IgG antibodies against tick-borne meningoencephalitis in both blood and cerebrospinal fluid. Despite the established intensive therapy, the patient further deteriorated and died on 27 August 2018 with a clinical picture of refractory failure of multiple organs. The suspicion of West Nile virus infection arose only post mortem, after consultation with an infectiologist; additional serological and molecular examinations then confirmed a WNV infection.

**Epidemiological background:** The first epidemiological inquiry was realized relatively soon after the patient´s death and focused primarily on environmental conditions at the place of residence. There were several barrels and tanks filled with stagnant water at the sites, which served as a suitable attractant for mosquitoes. Based on these findings, several $CO_2$ mosquito traps (Bioquip, Rancho Dominguez U.S.) were installed on 2 capture nights. A total of 87 female mosquitoes belonging to *Cx. pipiens*, *Anopheles maculipennis* sensu lato, *Aedes vexans*, *Ae. caspius* and *Ae. sticticus* were captured and examined for the presence of WNV. No positive sample was found. The patient did not leave the residence. Mosquitoes bit her regularly. She was not vaccinated against TBEV and did not undergo blood transfusion.

#### 2.1.3. Case 3

The 51-year-old man was admitted on 2 October 2018 to the Infectious Diseases Department of Břeclav Hospital for non-specific symptoms lasting from 9 September 2018—fever, headaches, fatigue, arthralgia and sleep disorders. This was a patient without significant comorbidities, who in the past was treated for Lyme disease and vertebrogenic algic syndrome. As part of the diagnosis, a lumbar puncture was added; biochemical and cytological tests of cerebrospinal fluid corresponded to aseptic neuroinfection. A CT scan of the brain showed no recent pathological changes. Tick-borne meningoencephalitis and herpesvirus or enterovirus neuroinfections were considered for differential diagnosis, but microbiological tests were negative for all these agents. The infectiologist also indicated that serological diagnosis of West Nile fever will be performed at the National Reference Laboratory (NRL) for arboviruses (Public Health Institute, Ostrava). Additional laboratory tests then confirmed a WNV infection. The clinical course of the disease was mild in this case; there were no complications during hospitalization, and the patient was discharged home after 14 days without a residual neurological deficit.

**Epidemiological background:** The epidemiological examination was performed soon after the patient returned from the hospital. The patient lives in a family house with garden and small natural pond; in some places there are barrels of stagnant water. He did not travel outside his residence, except for regular walks in the nearby forest. He denied vaccination against TBEV as well as blood donation/transfusion.

#### 2.1.4. Case 4

A 52-year-old man was admitted on 18 September 2018 to the Infectious Diseases Department of the Břeclav Hospital for non-specific symptoms lasting 8 days. At first, there were high fever, fatigue, arthralgia and myalgia, and the patient also reported headaches, diarrhea and vomiting during the past 4 days. The patient's medical history included arterial hypertension and long-term headaches of unclear origin. In the department of infectious diseases, lumbar puncture was added; biochemical and cytological tests of cerebrospinal fluid corresponded to aseptic neuroinfection. Due to the characteristics of the complaint, a CT scan of the brain was added with a finding of incipient cerebral oedema. However, later magnetic resonance imaging (MRI) of the brain did not show any intracranial inflammatory changes. The results of routine serological tests (including serology in tick-borne meningoencephalitis) were not convincing. In the light of previous experience (see Case reports 2 and 3), biological material was sent to the NRL for arboviruses for the diagnosis of West Nile fever. The laboratory also confirmed WNV infection in this case. In this case the hospitalization lasted 17 days, there were no complications, and the patient was discharged home in good condition.

**Epidemiological backgound:** Epidemiological investigation was realized by phone call. Patient lives in a family house with garden on the margin of town, close to a fishpond accompanied with rich vegetation. In the garden, several objects (barrels, watering cans) with stagnant water were present. He did not travel outside the region. He denied vaccination to TBEV as well as blood donation/transfusion.

#### 2.1.5. Case 5

A 46-year-old woman without a serious previous illness was admitted to the Infectious Diseases Clinic of Brno University Hospital on suspicion of meningoencephalitis on 10 January 2018. A 10-day tourist stay in Turkey preceded the symptoms, but the development of the symptoms occurred on the 3rd day after leaving for Turkey from the Czech Republic. First, a striking fatigue developed, with the development of headache in the following days and accompanied with nausea and vomiting, vertigo, and later the family observed bradypsichia and dysarthria. She had chills but did not measure her temperature. After admission, the patient was afebrile, stable in terms of the circulatory system, stupor, bradypsichia, still oriented in all qualities, with positive upper meningeal phenomena in the neurostat, fine tremor of the upper extremities and dysmetria, ataxia, standing titubation without lateral predilection. Other objective findings were remarkable (see Table 2). Only a slight increase in CRP was observed in the laboratory analysis; the other biochemical and haematological parameters were without deviations from the standard. On the day of the admission, a sample of cerebrospinal fluid was taken from the patient for examination—a picture of monocytic pleocytosis was present in the cerebrospinal fluid, the diagnosis of serous meningoencephalitis was confirmed in the context of the clinical condition. Anti-oedema therapy was administered to the patient—corticoids and osmotic diuretics—and a strict rest regime was indicated. Cerebrospinal fluid was sent for serological and molecular genetic testing. Herpes infection and enterovirus infections were eliminated by PCR. Tick-borne meningoencephalitis and Lyme borreliosis were also serologically excluded. After excluding other probable causes, the West Nile virus was suspected—cerebrospinal fluid, serum and urine were sent to the NRL for arboviruses for further analysis. The patient's health gradually improved during hospitalization, and anti-oedema doses decreased. As a complication of the corticosteroid therapy, the patient developed deep vein thrombosis of the right lower limb; given this result, anti-coagulant therapy was indicated. The complete serological results of the NRL examination were not absolutely unambiguous; admitted as a possible WNV pathogen, but Usutu virus infection was considered more likely. There is no causal treatment for any of these diseases, only symptomatic treatment is available; the patient was discharged from the hospital in good clinical condition on the 12th day of the hospitalization. At the follow-up visit on day 10 after discharge, the patient was completely free of symptoms; the neurological finding was negative.

**Epidemiological background**: Due to the development of symptoms soon after arrival in Turkey, the patient probably developed the disease before departure. The patient has a permanent residence in South Moravia and has reported mosquito biting.

### 2.2. Diagnostic Summary

Diagnostic outcomes are summarized in Table 3. The diagnosis of our patients was based on a case definition published by European Union [EU]. RealStar® WNV RT-PCR Kit 1.0 (Altona Diagnostics GmbH, Hamburg, Germany) assay was used for PCR and positive samples were sequenced [15]. Patient sera were tested for anti-WNV IgG and IgM with three commercial assays: Anti-West Nile virus IIFT IgG and IgM, Anti-West Nile virus ELISA IgG and IgM (both Euroimmun, Lübeck, Germany), ELISA West Nile Virus IgG and IgM capture Dx SelectTM (Focus Diagnostics, Cypress, CA, USA) and for anti-TBEV IgG and IgM with ELISA Viditest anti-TBEV IgG, IgG avidity and IgM (Vidia, Vestec, Czech Republic). VNT for WNV, USUV and TBEV according to a previously published internal protocol were used to confirm serological results [16,17].

In three patients both PCR and serology were positive, while one was not tested for PCR and one had positive antibodies but negative PCR. Only one of the WNV-positive samples (Patient 2) was confirmed to be a WNV-lineage 2 virus (by sequencing) and revealed a WNV-2 strain almost identical to WNV-2, which was detected in local mosquitoes [3]. Other samples could not be sequenced due to a low amount of viral RNA in samples. Therefore, USUV identification was not confirmed by sequencing or PCR. Positive serology was confirmed by a virus neutralization test (VNT) for WNV in all 5 patients; therefore, they met the laboratory criteria for a confirmed case. Subsequently, established VNT for USUV detected anti-USUV antibodies in 4 of the 5 previously diagnosed WNV patients. Three patients had anti-USUV VNT titers significantly lower than anti-WNV, presumably corresponding to cross-reactivity. The 5th patient had anti-USUV VNT titer 16 times higher than the anti-WNV, implying she was most likely to have a USUV infection, although her results originally matched the EU definition for WNV. Regarding serology, cross-reactivity was detected in the ELISA, while no patient had a equivocal positive anti-TBEV VNT.

## 3. Discussion

West Nile fever is now the most important mosquito-borne viral disease in Europe. The incidence of WNV peaked in 2018 with a total number of 2083 confirmed human cases (a 7.2-fold increase over the previous year) in Europe [18]. The massive outbreaks affected mainly Southern (Italy, Greece, Spain, France), Eastern (Croatia, Serbia, Bulgaria, Romania) and Central (Austria, Hungary, the Czech Republic) Europe, and expansion into previously virus-free regions (Slovenia, Kosovo). In a broader context, we should also consider the role of genetic, ecological, environmental and possible socio-economic aspects that may have played a role in increased WNV activity during the 2018 transmission season, most importantly suitable environmental factors for mosquito vectors, particularly increased day temperature [18] as well as more specific environmental factors such as a large number of vessels with stagnant water (barrels, watering cans and containers), which are constantly found in urban areas in the summer months (applies mainly to the Czech Republic). These objects represent ideal places for mass breeding of WNV vectors.

The Czech Republic and Germany are countries with the northernmost spread of WNV in Europe [3,19]. Increased surveillance, including large-scale surveys of mosquitoes, horses and birds, carried out during two large-scale EC-funded cooperation projects (EDEN and EDENext) between 2008 and 2015, has long indicated that WNV cases may occur in the country. As for the supervision of birds, Hubálek et al. [20] examined 54 domestic birds (geese and ducks) and 391 wild birds representing 28 migratory and resident species, using VNT in the South Moravian fishpond ecosystem. Antibodies to WNV were not detected in domestic waterfowl, but 23 (5.9%) wild birds of 10 species showed a positive response. Straková et al. [21] examined antibodies against WNV and USUV in 146 common coots (*Fulica atra*) on ponds in Moravia. Our results show that both WNV and USUV infections occur in common coots, and this species of bird can serve as an "indicator" of the presence of these viruses in fishpond and wetlands in Central Europe. In addition, two goshawks (*Accipiter gentilis*) held captive by falconers in Moravia died of WNV encephalitis in 2017 [22] and among predators, especially goshawks (several of them wild), WNV encephalitis broke out in the Czech Republic in 2018 [23].

According to recent data from several European countries [24,25], goshawks can serve as suitable indicators for active WNV circulation during the summer season in Europe. As far as horse surveillance is concerned, no case of West Nile fever has been reported in horses so far. The State Veterinary Institute in cooperation with the reference laboratory for arboviruses, regularly examines horse sera from all districts in the Czech Republic. Blood sera from 163 horses were examined from various parts of the Czech Republic in a plaque reduction neutralization test (VNT), but no specific WNV antibodies were detected [26]. A similar examination of a much larger sample of horses (2349 animals) revealed 11 horses (0.47%) with specific antibodies to WNV [27]. Regarding mosquito monitoring, WNV-2 was detected (RT-PCR) in *Culex modestus* mosquitoes collected in ponds in South Moravia during August 2013 and also isolated (newborn mice). Phylogenetic analysis has shown that these Czech WNV strains are closely related to the Austrian, Italian and Serbian strains reported in 2008, 2011 and 2012, respectively [3]. A total of 61,770 female *Cx. modestus* were collected in South Moravian ponds in the years 2010 to 2014, and 1243 samples were examined for the presence of flaviviruses by RT-PCR. Nine strains of WNV lineage 2 were detected in *Cx. modestus* collected in the same reed ecosystem. USUV and WNV co-circulate in the same wetland ecosystem, characterized by the presence of waterfowl and *Cx. modestus* mosquitoes, serving as hosts and vectors, respectively, for both viruses [8]. In addition, ornithophilic *Cx. pipiens* was demonstrated as a vector in 2015 [4]. A total of 28,287 hibernating mosquitoes caught in February or March from 2011 to 2017 in a WNV-endemic area of South Moravia were screened for the presence of WNV RNA. No WNV positive pools were found from 2011 to 2016, while lineage 2 WNV RNA was detected in 3 pools of *Cx. pipiens* mosquitoes collected in 2017 at 2 study sites. The data support the hypothesis of possible WNV persistence in mosquitoes throughout the winter season in Europe [28]. Interestingly, antibodies to WNV (overall 5.9% prevalence) were documented by VNT in the blood sera of wild artiodactyls including roe deer, red deer, fallow deer, mouflons and wild boars, sampled in the South Moravian district of Břeclav [29].

Blood safety testing started after the first WNV human cases were confirmed in the affected area (from September until November 2018). The Transfusion and Tissue Department of the University Hospital Brno started WNV testing of blood donors by PCR in September 2018 and finished at the end of November 2018. This solution was based on an epidemiological situation in South Moravia published on the websites of the European Centre for Disease Prevention and Control (ECDC). During this period, 4400 blood donors were tested with negative results.

## 4. Conclusions

In conclusion, our results confirm simultaneous circulation of WNV and USUV in the Czech Republic, so far limited to Southern Moravia. However, the first WNV-positive mosquitoes recently found in another region of the Czech Republic (Southern Bohemia) may indicate new WNV focus in the country [30]. Only a One-Health approach practicing interdisciplinary collaboration among local infection specialists, epidemiologists, veterinarians and entomologists, can bring benefit in the prevention and control of WNV in affected areas and also detect the introduction of WNV in previously virus-free areas in a timely manner.

**Author Contributions:** Conceptualization, H.Z., S.Š. and I.R.; funding acquisition, I.R.; investigation, H.Z., J.K., S.Š., P.S., H.K., R.S., P.H., P.H.J., E.T., H.L., O.Š., P.J., R.C., J.M. and I.R.; project administration, I.R.; supervision, H.Z. and I.R.; writing—original draft, H.Z., H.K., R.S., E.T., R.C. and I.R.; writing—review and editing, S.Š., P.S., R.S., E.T., H.L. and I.R. All authors have read and agreed to the published version of the manuscript.

**Funding:** The study was financially supported by the Ministry of Health of the Czech Republic (Reg. No. NV19-09-00036). All rights reserved. This work was also supported by the Ministry of Health, Czech Republic—conceptual development of research organization (FNBr, 65269705). Petra Strakova was supported by the Veterinary Research Institute (RVO: RO 0518).

**Institutional Review Board Statement:** Not applicable.

**Informed Consent Statement:** Patient consent was waived: strictly anonymous epidemiological, laboratory and clinical findings were applied in the study.

**Data Availability Statement:** Not applicable.

**Conflicts of Interest:** The authors declare no conflict of interest.

## References


1. Hubálek, Z.; Rudolf, I. *Microbial Zoonoses and Sapronoses*; Springer: Dordrecht, The Netherlands, 2011.
2. Hernández-Triana, L.M.; Jeffries, C.L.; Mansfield, K.L.; Carnell, G.; Fooks, A.R.; Johnson, N. Emergence of West Nile virus lineage 2 in Europe: A review on the introduction and spread of a mosquito-borne disease. *Front. Public Health* **2014**, *2*, 271. [CrossRef]
3. Rudolf, I.; Bakonyi, T.; Šebesta, O.; Peško, J.; Venclíková, K.; Mendel, J.; Betášová, L.; Blažejová, H.; Straková, P.; Nowotny, N.; et al. West Nile virus lineage 2 isolated from *Culex modestus* mosquitoes in the Czech Republic, 2013: Expansion of the European WNV endemic area to the North? *Eurosurveillance* **2014**, *19*, 20867. [CrossRef] [PubMed]
4. Rudolf, I.; Blažejová, H.; Šebesta, O.; Mendel, J.; Peško, J.; Betášová, L.; Straková, P.; Šikutová, S.; Hubálek, Z. West Nile virus (liniage 2) in mosquitoes in South Moravia—Awaiting first autochthonous human cases. *Epidemiol. Mikrobiol. Imunol.* **2018**, *67*, 44–46.
5. Weissenboeck, H.; Kolodziejek, J.; Url, A.; Lussy, H.; Rebel-Bauder, B.; Nowotny, N. Emergence of Usutu virus, an African mosquito-borne flavivirus of the Japanese encephalitis virus group, central Europe. *Emerg. Infect. Dis.* **2002**, *8*, 652–656. [CrossRef] [PubMed]
6. Hubálek, Z.; Rudolf, I.; Čapek, M.; Bakonyi, T.; Betášová, L.; Nowotny, N. Usutu Virus in Blackbirds (*Turdus merula*), Czech Republic, 2011–2012. *Transbound. Emerg. Dis.* **2014**, *61*, 273–276. [CrossRef]

7. Hönig, V.; Palus, M.; Kašpar, T.; Zemanová, M.; Majerová, K.; Hofmannová, L.; Papezik, P.; Šikutová, S.; Rettich, F.; Hubálek, Z.; et al. Multiple lineages of Usutu virus (*Flaviviridae, Flavivirus*) in blackbirds (*Turdus merula*) and mosquitoes (*Culex pipiens, Cx. modestus*) in the Czech Republic (2016–2019). *Microorganisms* **2019**, *7*, 568. [CrossRef] [PubMed]
8. Rudolf, I.; Bakonyi, T.; Šebesta, O.; Mendel, J.; Peško, J.; Betášová, L.; Blažejová, H.; Venclíková, K.; Straková, P.; Nowotny, N.; et al. Co-circulation of Usutu virus and West Nile virus in a reed bed ecosystem. *Parasit. Vectors* **2015**, *8*, 520. [CrossRef]
9. Cavrini, F.; Gaibani, P.; Longo, G.; Pierro, A.M.; Rossini, G.; Bonilauri, P.; Gerunda, G.E.; di Benedetto, F.; Pasetto, A.; Girardis, M.; et al. Usutu virus infection in a patient who underwent orthrotropic liver transplantation. *Eurosurveillance* **2009**, *14*, 19448. [PubMed]
10. Simoni, Y.; Sillam, O.; Carles, M.J.; Gutierrez, S.; Gil, P.; Constant, O.; Martin, M.F.; Girard, G.; Van de Perre, P.; Salinas, S.; et al. Human Usutu virus infection with atypical neurologic presentation, Montpellier, France, 2016. *Emerg. Infect. Dis.* **2018**, *24*, 875–878. [CrossRef]
11. Vilibic-Cavlek, T.; Kaic, B.; Barbic, L.; Pem-Novosel, I.; Slavic-Vrzic, V.; Lesnikar, V.; Kurecic-Filipovic, S.; Babic-Erceg, A.; Listes, E.; Stevanovic, V.; et al. First evidence of simultaneous occurrence of West Nile virus and Usutu virus neuroinvasive disease in humans in Croatia during the 2013 outbreak. *Infection* **2014**, *42*, 689–695. [CrossRef]
12. Gaibani, P.; Pierro, A.; Alicino, R.; Rossini, G.; Cavrini, F.; Landini, M.P.; Sambri, V. Detection of Usutu-virus-specific IgG in blood donors from Northern Italy. *Vector Borne Zoonotic Dis.* **2012**, *12*, 431–433. [CrossRef] [PubMed]
13. Bakonyi, T.; Jungbauer, C.; Aberle, S.W.; Kolodziejek, J.; Dimmel, K.; Stiasny, K.; Allerberger, F.; Nowotny, N. Usutu virus infections among blood donors, Austria, July and August 2017—Raising awareness for diagnostic challenges. *Eurosurveillance* **2017**, *22*, 41. [CrossRef] [PubMed]
14. Allering, L.; Jost, H.; Emmerich, P.; Gunther, S.; Lattwein, E.; Schmidt, M.; Seifried, E.; Sambri, V.; Hourfar, K.; Schmidt-Chanasit, J. Detection of Usutu virus infection in a healthy blood donor from southwest Germany. *Eurosurveillance* **2012**, *17*, 20341. [CrossRef] [PubMed]
15. Scaramozzino, N.; Crance, J.M.; Jouan, A.; DeBriel, D.A.; Stoll, F.; Garin, D. Comparison of flavivirus universal primer pairs and development of a rapid, highly sensitive heminested reverse transcription-PCR assay for detection of flaviviruses targeted to a conserved region of the NS5 gene sequences. *J. Clin. Microbiol.* **2001**, *39*, 1922–1927. [CrossRef] [PubMed]
16. Zelená, H.; Januška, J.; Raszka, J. Micromodification of virus-neutralisation assay with vital staining in 96-well plate and its use in diagnostics of Ťahyňa virus infections. *Epidemiol. Mikrobiol. Imunol.* **2008**, *57*, 104–108.
17. Litzba, N.; Zelená, H.; Kreil, T.R.; Niklasson, B.; Kühlmann-Rabens, I.; Remoli, M.E.; Niedrig, M. Evaluation of different serological diagnostic methods for tick-borne encephalitis virus: Enzyme-linked immunosorbent, immunofluorescence, and neutralization assay. *Vector Borne Zoonotic Dis.* **2014**, *14*, 1–113. [CrossRef]
18. Nowotny, N.; Camp, J.V. The knowns and unknowns of West Nile virus in Europe: What did we learn from the 2018 outbreak? *Exp. Rev. Anti Infect. Ther.* **2020**, *18*, 145–154.
19. Pietsch, C.; Michalski, D.; Munch, J.; Petros, S.; Bergs, S.; Trawinski, H.; Lubbert, C.; Liebert, U.G. Autochthonous West Nile virus infection outbreak in humans, Leipzig, Germany, August to September 2020. *Eurosurveillance* **2020**, *25*, 2001786. [CrossRef]
20. Hubálek, Z.; Halouzka, J.; Juřicová, Z.; Šikutová, S.; Rudolf, I.; Honza, M.; Janková, J.; Chytil, J.; Marec, F.; Sitko, J. Serologic survey of birds for West Nile flavivirus in southern Moravia (Czech Republic). *Vector Borne Zoonotic Dis.* **2008**, *8*, 659–666. [CrossRef]
21. Straková, P.; Šikutová, S.; Jedličková, P.; Sitko, J.; Rudolf, I.; Hubálek, Z. The Common Coot as sentinel species for the presence of West Nile and Usutu flaviviruses in Central Europe. *Res. Vet. Sci.* **2015**, *102*, 159–161. [CrossRef]
22. Hubálek, Z.; Kosina, M.; Rudolf, I.; Mendel, J.; Straková, P.; Tomešek, M. Mortality of goshawks (*Accipiter gentilis*) due to West Nile virus lineage 2. *Vector Borne Zoonotic Dis.* **2018**, *18*, 624–627. [CrossRef]
23. Hubálek, Z.; Tomešek, M.; Kosina, M.; Šikutová, S.; Straková, P.; Rudolf, I. West Nile virus outbreak in captive and wild raptors, Czech Republic, 2018. *Zoonoses Public Health* **2019**, *66*, 978–981. [CrossRef]
24. Feyer, S.; Bartenschlager, F.; Bertram, C.A.; Ziegler, U.; Fast, C.; Klopfleisch, R.; Muller, K. Clinical, pathological and virological aspects of fatal West Nile virus infections in ten free-ranging goshawks (*Accipiter gentilis*) in Germany. *Transbound. Emerg. Dis.* **2020**, *68*, 907–919. [CrossRef]
25. Vidana, B.; Busquets, N.; Napp, S.; Perez-Ramirez, E.; Jimenez-Clavero, M.A.; Johnson, N. The Role of Birds of Prey in West Nile Virus Epidemiology. *Vaccines* **2020**, *8*, 550. [CrossRef]
26. Hubálek, Z.; Ludvíková, E.; Jahn, P.; Treml, F.; Rudolf, I.; Svobodová, P.; Šikutová, S.; Betášová, L.; Bíreš, J.; Mojžíš, M.; et al. West Nile virus equine serosurvey in the Czech and Slovak Republics. *Vector Borne Zoonotic Dis.* **2013**, *13*, 733–738. [CrossRef]
27. Sedlák, K.; Zelená, H.; Křivda, V.; Šatrán, P. Surveillance of West Nile fever in horses in the Czech Republic from 2011 to 2013. *Epidemiol. Mikrobiol. Imunol.* **2014**, *63*, 307–311.
28. Rudolf, I.; Betášová, L.; Blažejová, H.; Venclíková, K.; Straková, P.; Šebesta, O.; Mendel, J.; Bakonyi, T.; Schaffner, F.; Nowotny, N.; et al. West Nile virus in overwintering mosquitoes, Czech Republic. *Parasit. Vectors* **2017**, *10*, 452. [CrossRef] [PubMed]
29. Hubálek, Z.; Juřicová, Z.; Straková, P.; Blažejová, H.; Betášová, L.; Rudolf, I. Serological survey for West Nile virus in wild artiodactyls, Southern Moravia (Czech Republic). *Vector Borne Zoonotic Dis.* **2017**, *17*, 654–657. [CrossRef] [PubMed]
30. Rudolf, I.; Rettich, F.; Betášová, L.; Imrichová, K.; Mendel, J.; Hubálek, Z.; Šikutová, S. West Nile virus (lineage 2) detected for the first time in mosquitoes in Southern Bohemia: New WNV endemic area? *Epidemiol. Mikrobiol. Imunol.* **2019**, *68*, 150–153. [PubMed]

*pathogens*

MDPI

*Article*

# West Nile Virus in Brazil

Érica Azevedo Costa [1,†], Marta Giovanetti [2,3,†], Lilian Silva Catenacci [4,†], Vagner Fonseca [3,5,6,†], Flávia Figueira Aburjaile [3,†], Flávia L. L. Chalhoub [2], Joilson Xavier [3], Felipe Campos de Melo Iani [7], Marcelo Adriano da Cunha e Silva Vieira [8], Danielle Freitas Henriques [9], Daniele Barbosa de Almeida Medeiros [9], Maria Isabel Maldonado Coelho Guedes [1], Beatriz Senra Álvares da Silva Santos [1], Aila Solimar Gonçalves Silva [1], Renata de Pino Albuquerque Maranhão [10], Nieli Rodrigues da Costa Faria [2], Renata Farinelli de Siqueira [11], Tulio de Oliveira [5], Karina Ribeiro Leite Jardim Cavalcante [12], Noely Fabiana Oliveira de Moura [12], Alessandro Pecego Martins Romano [12], Carlos F. Campelo de Albuquerque [13], Lauro César Soares Feitosa [14], José Joffre Martins Bayeux [15], Raffaella Bertoni Cavalcanti Teixeira [16], Osmaikon Lisboa Lobato [17], Silvokleio da Costa Silva [17], Ana Maria Bispo de Filippis [2], Rivaldo Venâncio da Cunha [18], José Lourenço [19] and Luiz Carlos Junior Alcantara [2,3,*]

1 Departamento de Medicina Veterinária Preventiva, Universidade Federal de Minas Gerais, Belo Horizonte 31270-901, Brazil; azevedoec@yahoo.com.br (É.A.C.); mariaisabel.guedes@gmail.com (M.I.M.C.G.); beatrizsenra.santos@gmail.com (B.S.Á.d.S.S.); ailavet@yahoo.com.br (A.S.G.S.)

2 Laboratório de Flavivírus, Instituto Oswaldo Cruz, Fundação Oswaldo Cruz, Rio de Janeiro 21040-360, Brazil; giovanetti.marta@gmail.com (M.G.); flaviallevy@yahoo.com.br (F.L.L.C.); nielircf@gmail.com (N.R.d.C.F.); ana.bispo@ioc.fiocruz.br (A.M.B.d.F.)

3 Laboratório de Genética Celular e Molecular, Universidade Federal de Minas Gerais, Belo Horizonte 31270-901, Brazil; vagner.fonseca@gmail.com (V.F.); faburjaile@gmail.com (F.F.A.); joilsonxavier@live.com (J.X.)

4 Departamento De Morfofisiologia Veterinária, Universidade Federal do Piauí, Teresina 64049-550, Brazil; catenacci@ufpi.edu.br

5 KwaZulu-Natal Research Innovation and Sequencing Platform (KRISP), School of Laboratory Medicine and Medical Sciences, College of Health Sciences, University of KwaZulu-Natal, Durban 4041, South Africa; tulioDNA@gmail.com

6 Coordenação Geral dos Laboratórios de Saúde Pública/Secretaria de Vigilância em Saúde, Ministério da Saúde (CGLAB/SVS-MS), Brasília 70719-040, Brazil

7 Laboratório Central de Saúde Pública, Fundação Ezequiel Dias, Belo Horizonte 30510-010, Brazil; felipeemrede@gmail.com

8 Diretoria de Vigilância em Saúde, Fundação Municipal de Saúde, Teresina 64600-000, Brazil; marceloadrianoneuro@gmail.com

9 Seção de Arbovirologia e Febres Hemorrágicas, Instituto Evandro Chagas, Ministério da Saúde, Ananindeua 70058-900, Brazil; dannifh@hotmail.com (D.F.H.); danielemedeiros@iec.pa.gov.br (D.B.d.A.M.)

10 Setor de Clínica de Equinos, Hospital Veterinário, Campus Pampulha, Universidade Federal de Minas Gerais Escola de Veterinária, Belo Horizonte 31270-901, Brazil; rpamaranhao@yahoo.com

11 Department of Large Animal Clinic, Universidade Federal de Santa Maria, Rio Grande do Sul 97105-900, Brazil; renata.farinelli@ufsm.br

12 Coordenacao Geral das Arboviroses, Secretaria de Vigilância em Saúde/Ministério da Saúde, Brasília 70058-900, Brazil; karina.cavalcante@saude.gov.br (K.R.L.J.C.); noely.moura@saude.gov.br (N.F.O.d.M.); alessandro.romano@saude.gov.br (A.P.M.R.)

13 Organização Pan-Americana da Saúde, Organização Mundial da Saúde, Brasília 40010-010, Brazil; meloc@paho.org

14 Centro de Ciências Agrárias, Departamento de Clínica e Cirurgia Veterinária, Universidade Federal do Piauí, Teresina 64049-550, Brazil; jackvet08@hotmail.com

15 Faculdade de Ciências da Saúde, Medicina Veterinária, Urbanova, São José Dos Campos, UNIVAP-Universidade Vale do Paraíba, São Paulo 12245-720, Brazil; jjveterinario@hotmail.com

16 Departamento de Clínica e Cirurgia Veterinárias, Escola de Veterinária, Universidade Federal de Minas Gerais, Belo Horizonte 31270-901, Brazil; teixeiraraffa@gmail.com

17 Laboratório de Genética e Conservação de Germoplasma, Campus Prof. Cinobelina Elvas, Universidade Federal do Piauí, Bom Jesus, Piauí 64049-550, Brazil; osmaikonlobato@gmail.com (O.L.L.); silvokleio@ufpi.edu.br (S.d.C.S.)

18 Coordenacao dos Laboratorios de Referencia, Oswaldo Cruz Foundation, Rio de Janeiro 21040-360, Brazil; rivaldo.cunha@fiocruz.br

19 Department of Zoology, University of Oxford, Oxford OX1 3PS, UK; jose.lourenco@zoo.ox.ac.uk

* Correspondence: luiz.alcantara@ioc.fiocruz.br

**Citation:** Costa, É.A.; Giovanetti, M.; Silva Catenacci, L.; Fonseca, V.; Aburjaile, F.F.; Chalhoub, F.L.L.; Xavier, J.; Campos de Melo Iani, F.; da Cunha e Silva Vieira, M.A.; Freitas Henriques, D.; et al. West Nile Virus in Brazil. *Pathogens* **2021**, *10*, 896. https://doi.org/10.3390/pathogens10070896

Academic Editor: Francisco Llorente

Received: 30 April 2021
Accepted: 21 May 2021
Published: 15 July 2021

† Denote equal contribution.

**Abstract:** *Background:* West Nile virus (WNV) was first sequenced in Brazil in 2019, when it was isolated from a horse in the Espírito Santo state. Despite multiple studies reporting serological evidence suggestive of past circulation since 2004, WNV remains a low priority for surveillance and public health, such that much is still unknown about its genomic diversity, evolution, and transmission in the country. *Methods:* A combination of diagnostic assays, nanopore sequencing, phylogenetic inference, and epidemiological modeling are here used to provide a holistic overview of what is known about WNV in Brazil. *Results:* We report new genetic evidence of WNV circulation in southern (Minas Gerais, São Paulo) and northeastern (Piauí) states isolated from equine red blood cells. A novel, climate-informed theoretical perspective of the potential transmission of WNV across the country highlights the state of Piauí as particularly relevant for WNV epidemiology in Brazil, although it does not reject possible circulation in other states. *Conclusion:* Our output demonstrates the scarceness of existing data, and that although there is sufficient evidence for the circulation and persistence of the virus, much is still unknown on its local evolution, epidemiology, and activity. We advocate for a shift to active surveillance, to ensure adequate preparedness for future epidemics with spill-over potential to humans.

**Keywords:** West Nile virus; genomic monitoring; molecular detection; Brazil

## 1. Introduction

West Nile virus (WNV), a member of the *Flaviviridae* family, was first identified in the West Nile district of Uganda in 1937, but nowadays, it is commonly found in Africa, Europe, North America, the Middle East, and Asia [1–3]. WNV transmission is maintained in a mosquito–bird cycle, for which the genus *Culex*, in particular *Cx. pipiens* and *quinquefasciatus*, are considered the principal vectors [4]. WNV can infect humans, equines, and other mammals, but these are considered "dead-end" hosts, given their weak potential to function as amplifying hosts to spread infection onwards [5,6]. Around 80% of WNV infections in humans are asymptomatic, while the rest may develop mild or severe disease. Mild disease includes fever, headache, tiredness, and vomiting [7,8], while severe disease (neuroinvasive) is characterized by high fever, coma, convulsions, and paralysis [7,8]. Equine infections can occasionally cause neurological disease and death [7,8], such that equines typically serve as sentinel species for WNV outbreaks with potential for spill-over into human populations.

Genome detection of WNV in South America was originally reported in horses (Argentina in 2006) and captive flamingos (Colombia, in 2012) [9,10]. The first ever sequenced genome in Brazil was in 2018, when the virus was isolated from a horse with severe neurological disease in the Espírito Santo state [11]. Despite multiple studies reporting serological evidence suggestive of past WNV circulation in Brazil (e.g., [11–13]) and reports of human WNV disease in confirmed cases in the Piauí state [13], much is unknown about genomic diversity, evolution, and transmission dynamics across the country. The reality of WNV in Brazil is likely characterised by endemic circulation within the mosquito–bird cycle [14–17], with occasional transmission to humans. The so far lack of reported human epidemics with significant public health impact remains a puzzle, given that Brazil harbors the necessary vectors, avian species, and climate—combination amenable at sustaining endemicity [18]. Several factors potentially contribute to the seemingly silent circulation of WNV in the country [19], such as the lack of surveillance interest and resources, rates of mild human WNV disease, co-circulation of other mosquito-borne viruses that cause similar clinical spectrums, and diagnostics and screening of animals and humans well past the time of infection, which critically hampers viral detection and confirmation.

In this study, we aim at providing a holistic perspective of what is known about WNV circulation in Brazil. In addition to previously reported evidence of WNV circulation, we also report new genetic evidence of WNV circulation in three Brazilian states. We further provide a climate-informed, theoretical assessment of the transmission potential of WNV across Brazil, revealing spatio-temporal patterns of interest. The lack of surveillance data hampers more in-depth analyses and therefore obscures our current understanding of WNV epidemiology, evolution, and transmission in the country. Recently, some European countries have witnessed a shift from a similar surveillance and epidemiological situation to that of Brazil, to observing recurrent WNV epidemics with spill-over to human populations [18–21]. We argue that active surveillance initiatives are necessary in Brazil in the near future to ensure preparedness of future WNV epidemics with public health impact.

## 2. Results

### *2.1. Novel Evidence of WNV Circulation in Three Brazilian States*

Samples (RBCs) from three horses with suspected WNV infection obtained from southern (Minas Gerais and São Paulo) and northeastern (Piauí) Brazilian states were sent for molecular diagnosis at the Departamento de Medicina Veterinária Preventiva at the Federal University of Minas Gerais (UFMG).

RNAs were extracted from red blood cells and tested using an in-house PCR assay (see Methods section for details). WNV-specific RT-PCR amplification products were obtained by nested PCR (Figure 1A,B), and positive samples were subjected to a newly designer multiplex PCR scheme (Supplementary Table S1) to generate complete genomes sequences by means of portable nanopore sequencing.

Three blood fractions (plasma, buffy coat, and washed RBC) from the horses sampled in São Paulo and Minas Gerais states have been submitted to nested RT-PCR; horse samples from Piaui have been tested only using RBC, which was the only blood fraction available. Diagnostic investigation of alphavirus was also performed using a generic RT-PCR targeting the NSP1 gene, according to [22], in the three blood fractions, with negative results.

The published WNV genome from Brazil (MH643887) was used to generate (mean) 98.4% consensus sequences that formed the target for primer design. The new genomes were deposited in GenBank with accession numbers MW420987, MW420988, and MW420989 (Table 1).

We constructed phylogenetic trees to explore the relationship of the sequenced genomes to those sampled elsewhere globally. We retrieved 2321 WNV genome sequences with associated lineage date and country of collection from GenBank, from which we generated a subset that included the highly supported (>0.9) clade containing the newly WNV strains obtained in this study plus 29 sequences (randomly sampled) from all lineages and performed phylogenetic analysis. An automated online phylogenetic tool to identify and classify WNV sequences was developed (available at: http://krisp.ukzn.ac.za/app/typingtool/wnv/job/9b40f631-51c4-419c-9edf-2206e7cd8d9c/interactive-tree/phylo-WNV.xml accessed on 31 December 2019).

Phylogenies estimated by the newly developed WNV typing tool, along with maximum likelihood methods (Supplementary Figure S1C), consistently placed the Brazilian genomes in a single clade within the 1a lineage with maximum statistical support (bootstrap = 100%) (Supplementary Figure S1).

Time-resolved maximum likelihood tree appeared to be consistent with previous estimates [11] and showed that the new genomes clustered with strong bootstrap support (97%) with a WNV strain isolated from an *Aedes albopictus* mosquito in Washington DC, USA in 2019 (Figure 1D). Interestingly, the new isolates did not group with the previously sequenced genome in 2019 from the Espirito Santo state, suggesting that inter-continental introduction events might be frequent in Brazil.

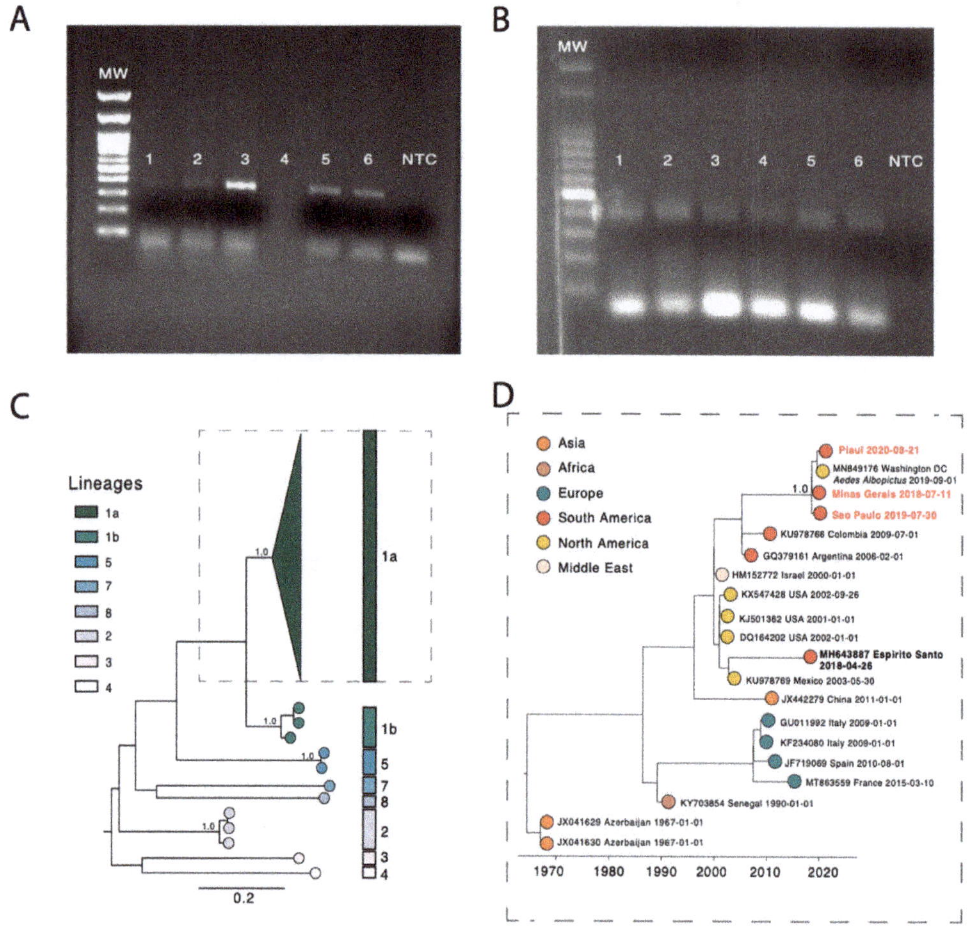


**Figure 1.** Investigation of WNV infections in Brazil, between July 2018 and September 2020, and estimated transmission potential. (**A**,**B**) Agarose gel electrophoresis of amplicons from assay for WNV. (**A**) nested RT-PCR. MW (Molecular weight ladder), 100 bp DNA Ladder RTU, Kasvi; 1—plasma of horse from São Paulo; 2—buffy coat of horse from São Paulo; 3—washed RBC of horse from São Paulo; 4—blank negative control using during the nested RT-PCR; 5 and 6—positive control (synthetic gene); NTC, no template control (using since the extraction); expected amplicon size: 370 bp. (**B**) Multiplex PCR. MW (Molecular weight ladder), Fluorescent 100 bp DNA Ladder, Cellco, Jena Bioscience; 1—horse form Minas Gerais (pair primers); 2—horse form Minas Gerais (odd primers); 3—horse form Sao Paulo (pair primers); 4—horse form Sao Paulo (impair primers); 5—horse form Piaui (pair primers); 6—horse form Piaui (odd primers); NTC, no template control (using since the extraction); expected amplicon size: 400 bp. (**C**) Midpoint rooted maximum-likelihood phylogeny of WNV genomes, showing major lineages. The scale bar is in units of substitutions per site (*s/s*). Support for branching structure is shown by bootstrap values at nodes. (**D**) Time-resolved maximum likelihood tree showing the WNV strains belonged to the 1a lineage. Colors indicate geographic location of sampling. The new Brazilian WNV strains are shown with text in red.

**Table 1.** Epidemiological information and sequencing statistics of the three sequenced samples of WNV sampled in Minas Gerais, São Paulo, and Piaui Brazilian states.

| ID | Sample | Collection Date | Age | Sex | State | Municipality | Reads | Coverage (%) | Depth of Coverage | Lineage Assignment | Acession Number | Clinical Sign |
|---|---|---|---|---|---|---|---|---|---|---|---|---|
| BC02_07 | RBCs | 11/07/2018 | 9 months | F | MG | Sabara | 343,743 | 97.9 | 6527.6 | Lineage 1a | MW420989 | Chorioretinitis |
| BC03_04 | RBCs | 30/07/2019 | 13 years-old | M | SP | São Bernardo do Campo | 170,980 | 97.9 | 3189.7 | Lineage 1a | MW420988 | Muscle stiffness, tremor retinal and flaccid paralysis |
| BC05_06 | RBCs | 21/08/2020 | 5 years-old | F | PI | Parnaíba | 222,516 | 99.4 | 4121.4 | Lineage 1a | MW420987 | Neurological complications |

ID = study identifier; RBCs = Red Blood Cells; Collection date = Sample collection date; Municipality = Municipality of residence; State= MG-Minas Gerais; SP = Sao Paulo; PI = Piaui; Sex: M = Male; F = Female; Accession Number = NCBI accession number.

### 2.2. *A Data-Driven WNV Theoretical Perspective*

We first summarized the past evidence of WNV circulation in Brazil from avian species, equines, and humans, which was achieved via various literature reports using different confirmation methods (Figure 2A) [23]. The first evidence of WNV infection was documented in 2004 in horses in northeastern Brazil (Paraiba state). Since then, serological evidence of WNV infection continued to be documented between 2008 and 2010 and again in 2020 in horses and birds from the southern [24], midwestern (Pantanal), and northern Brazilian regions. In 2014, the first WNV infection in a human was confirmed in the Piauí State (northeast region). In 2018, the first isolation of WNV in Brazil was documented in the Espirito Santo state (southeastern Brazil) when the virus was isolated from the central nervous system (CNS) of a dead horse with neurological manifestations [11]. To these data, we here add the report of the new genetic evidence of WNV circulation in equines occurring between 2018 and 2020, in southern (Minas Gerais, São Paulo) and northeastern (Piauí) states. To the best of our knowledge, it is the first time that evidence of WNV circulation is reported for the states of Minas Gerais and São Paulo.

Using data collected from the Brazilian "Sistema de Informação de Agravos de Notificação" (SINAN) (see Methods and Supplementary Table S2) reported with identifier A923 ("Febre do Nilo"), we explored the current spatio-temporal distribution of suspected cases of West Nile fever. Given the unspecific and unconfirmed nature of these reported cases, we complemented such information with theoretical projections of the spatio-temporal transmission potential of WNV in Brazil. For this, we used a climate-driven suitability measure (index P) previously successfully applied to WNV in the contexts of Israel [25] and Portugal [26] (see Methods).

We mapped the mean index P across Brazil for the period 2015–2019 (Figure 2B) and found estimated transmission potential to be highest in the center of the country along a diagonal latitude–longitude axis crossing from the center–west to the north–east. The regions of the south of the country, similarly to estimations for other mosquito-borne viruses [27], presented the least transmission potential. To assess potential hotspots of (at least temporary) high transmission potential, we calculated the proportion of months (2015–2019) in which the index P was above 1; this particular threshold representing the point above which each female mosquitoes would be theoretically able to infect more than one host during their lifetime. This approach identified regions of Piauí, Bahia, Ceará, Rio Grande do Norte, and Paraíba states as presenting significantly longer periods of time with high index P. In particular, the state of Piauí was captured in its entirety within this estimated spatial hotspot of transmission potential (Figure 2C).

From all states for which there were reported cases, we filtered those that had more than one case per any month during the entire period of 2015–2019, selecting only two states with clear epidemic waves of reported cases: Piauí and Espírito Santo. Coincidently with the results of Figure 2B,C, the state of Piauí reported the largest number of cases in the entire dataset. Using the geographical boundaries of each state, we averaged the index P per month (Figure 2D,E). The resulting time series of transmission potential showed that potential was higher in Piauí compared to Espírito Santo in accordance with the spatial output in Figure 2B,C. It also presented a clear seasonal signal, with peaks occurring on average in February in Piauí (month average = 2.2, summer) and April in Espiríto Santo (month average = 4, autumn). The correlation between reported cases and the index for Piauí was positive (Pearson's 0.36, Figure 2D), but it was negative for Espirito Santo (Pearson's −0.31, Figure 2E). Similar to what has been reported for suitability indices applied to other viruses [27], there was a clear lag between the index and cases for Piauí, with cases lagging behind the index (Figure 2D). Accordingly, shifting the index by one month into the future resulted in a high positive correlation with cases (Pearson's 0.84).

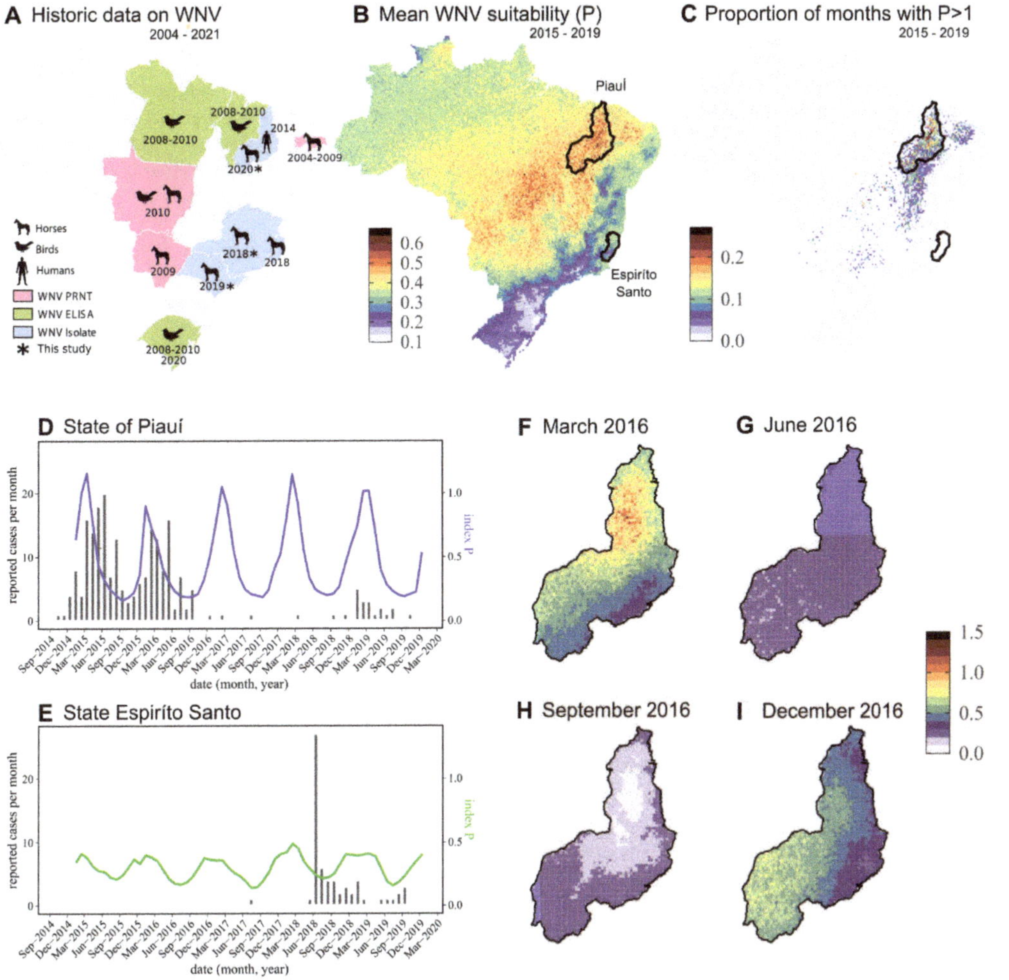


**Figure 2.** Data-driven epidemiological perspective of WNV in Brazil. (**A**) Mapping of historic evidence for WNV circulation in Brazil, for which the color and symbol legend on the bottom left of the panel define the animal source and methodology. Data are based on a literature review up to 2019 [24], in addition with recently published reports in 2020–2021 [24] and the new data generated in this study. (**B**) Mean estimated transmission potential of WNV (index P) over the period 2015–2019. The color scale on the bottom left of the panel shows the range of the presented values. The black borders mark the boundaries of the Piauí and Espiríto Santo states. (**C**) Proportion of months for which the transmission potential of WNV (index P) was above the value 1, over the period 2015–2019. The color scale on the bottom left of the panel shows the range of the presented values. The black borders mark the boundaries of the Piauí and Espiríto Santo states. (**D**) Time series of suspected reported West Nile fever cases (bars) and estimated transmission potential of WNV (index P, blue line) for the Piauí state. Index P is the average per month, across all data points within the boundaries of the state. (**E**) Time series of suspected reported West Nile fever cases (bars) and estimated transmission potential of WNV (index P, green line) for the Espiríto Santo state. Index P is the average per month, across all data points within the boundaries of the state. (**F**) Spatial snapshot of estimated transmission potential of WNV (index P) for the month of March 2016. Color scale on the right shows the range of the presented values. (**G**) Same as F but for June 2016. (**H**) Same as F but for September 2016. (**I**) Same as F but for December 2016.

Finally, to get a grasp of the possible spatio-temporal dynamics of WNV transmission in Piauí, we looked at estimated transmission potential for one of the years with more reported suspected cases (2016) both in space and time (with snapshots at months of March, June, September, and December) (Figure 2F–I). The spatio-temporal snapshots showed that transmission potential was the lowest during winter months, but we also highlighted that this was almost uniform across the state (Figure 2G,H). In contrast, throughout the year, this output highlighted a possible wave of seasonal transmission. This wave would typically start in the southwest just before summer (Figure 2I) and would move to the northeast in the summer (Figure 2F).

## 3. Discussion

Our analyses indicate that additional data are required to better identify routes of WNV importation into and within Brazil and to more generally understand the local transmission dynamics of the virus. Interestingly, our data suggest that the circulation of the virus may have resulted from multiple independent introductions, since the new isolates did not group with the previously sequenced genome in 2019 from the Espirito Santo state. This suggests that intra-continental introduction events due to the mobility of infected birds or mosquitoes might be a more plausible mechanism for the multiple introductions of WNV in South American countries, including Brazil. This scenario is consistent with previous studies that showed that multiple independent introductions into Latin America occurred during the initial outbreak in US in 1999; detailed revision is provided in [28]. While migrating birds are a convenient explanation of WNV dispersal, other possible ways of dispersion exist, such as infected mosquitoes that are accidentally transported via airplane or by road transport [29]. Another likely scenario is commercial legal or ilegal human transportation of birds and/or mosquitoes, which could be transported on airplanes [29].

The current data scarceness prevents definite conclusions on key aspects of WNV epidemiology. For example, given the unconfirmed nature of the reported cases by SINAN for Piauí and Espírito Santo, it is unclear what the proportion of cases truly reflect WNV occurrence and seasonality, hampering our ability to ascertain how representative our theoretical projections are. For Piauí, we would speculate that reported cases may indeed reflect some aspects of WNV seasonality, given that this state had the largest number of cases reported while also being the region of Brazil for which we estimated higher transmission potential and that our estimated transmission potential was well correlated with reported cases (albeit with a possible lag of one month typical of mosquito-borne viruses). At the same time, while inferred trees including the new genome sequences suggest that inter-continental introduction events might be frequent in Brazil, the lack of higher spatio-temporal sampling restricts our ability for definite conclusions on viral movement and persistence.

The phylogenetic and epidemiologic perspectives presented in this study, based on both existing and novel data as well as theoretical projections, suggest that both scenarios of sporadic and endemic local transmission are possible [30]. Similarly to sudden changes in WNV epidemiology and transmission as recently observed in other countries, the occurrence of a WNV outbreak affecting humans in Brazil may simply be a matter of time. Shifting from passive to active WNV screening and sequencing in animal reservoirs (e.g., equines, birds, vectors) in Brazil must be implemented to better understand the virus' local epidemiology and to be able to act accordingly in preventing and controlling any future epidemics with spill-over to humans.

## 4. Materials and Methods

### 4.1. Sample Collection, Viral RNA Isolation and PCR Screening

Samples (red blood cells, RBCs) from three horses with suspected WNV infection obtained from southern (Minas Gerais and São Paulo) and northeastern (Piauí) Brazilian states were sent for molecular diagnosis at the Laboratório de Patologia Molecular at the Federal University of Minas Gerais (UFMG).

Sample 1 from 11 July 2018 was collected from a 9-month-old female horse in a farm in the state of Minas Gerais, Mangueiras neighbourhood (Sabará), 15 km from the capital Belo Horizonte. Clinical findings were consistent with bilateral blindness. Neurological examination revealed no other abnormalities. The ophthalmological exams (direct and indirect pupillary light reflex (PLR), fluorescein eye stain test, fundus examination, and intraocular pressure) were consistent with retinal disease, mainly with chorioretinitis.

Sample 2 from 30 July 2019 was collected from a 13-year-old male horse that presented seizure episodes, muscle stiffness, tremor retinal, and flaccid paralysis in a farm located in São Bernardo do Campo countryside of the São Paulo state. Twenty-four days after the onset of neurological signs, the animal had severe pain in the forelimbs from laminitis, and it was euthanized due to hoof decumulation.

Sample 3 from 21 August 2020 was collected from a male horse, 5 years old, which died 72 h after presenting neurological signs, in a farm located in the municipality of Parnaíba, Piauí state. The animal presented motor incoordination, paddling movements, loss of sensitivity over the spine column, and behavioral changes. In this municipality, the tenth human case in Brazil was also detected, presenting neuroinvasive disease compatible with WNV infection, confirmed by serological assay (IgM) in both serum and cerebrospinal fluid (CSF) samples during acute and convalescent phases.

Whole blood samples obtained from the three horses were centrifuged at 1260× *g* for 20 min, and the plasma and buffy coat fractions were collected and stored at 4 °C. Red blood cells (RBC) were washed by centrifugation three times in phosphate-buffered saline (PBS) at 1260× *g* for 10 min and stored also at 4 °C [15]. RNA from each unit (washed RBC, plasma and buffy coat) were extracted using the QIAmp Viral RNA Mini kit (Qiagen, Hilden, Germany), following manufacturer's recommendations.

Diagnostic investigation of arboviruses was performed by a generic RT-PCR targeting the flavivirus non-structural protein 5 (NS5) gene [31] and alphavirus non-structural protein 1 gene (nsP1) [32]. West Nile virus-specific degenerated primers: forward primers (+) AACCKCCAGAAGGAGTSAAR and reverse primers (−) AGCYTCRAACTCCAGRAAGC were used in second reaction of nested PCR targeting the NS5 gene after a genus specific flavivirus RT-PCR amplification [22]. A synthetic gene fragment of partial NS5 gene (gblocks gene fragment, Integrated DNA Technologies) was used as a positive control. The 25 μL PCR "master-mix" comprised 2.5 μL of 10× PCR buffer, 1.5 mM MgCl2, 0.4 μM of each primer (forward and reverse), 0.8 μM dNTP mixture (Phoneutria, Sao Paulo, Brazil), 1 U Taq DNA polymerase (Platinum Taq DNA polymerase; Invitrogen, Carlsbad, CA, USA), 2 μL of template DNA (sample or gBlock), and DNA/RNAse-free water. The thermocycling conditions involved 40 cycles, and reaction conditions were previously reported in [18]. As an internal control for amplification efficiency, primers for the beta actin gene were used. As a negative control for the reactions, we used RNA extracted from equine washed RBC, plasma, and buffy coat that previously tested negative for arboviruses, equine herpesvirus 1 and 4, and borna disease. The amplicons were analyzed by 1% (*w/v*) agarose gel electrophoresis, stained with ethidium bromide, and visualized under UV light. Nested PCR were performed for equine herpesvirus 1 (EHV-1) [33] for borna disease [34,35], both with negative results in the 3 horses.

### 4.2. cDNA Synthesis and Multiplex Tiling PCR

Then, WNV-positive (in nested RT-PCR) RNA samples from washed RBCs were submitted to a cDNA synthesis protocol [36] using a Superscript IV cDNA Synthesis Kit. Then, a multiplex PCR primer scheme was designed (Table S1) to generate complete genomes sequences by means of portable nanopore sequencing, using Primal Scheme (Supplementary Table S1) (http://primal.zibraproject.org accessed on 31 December 2019) [37]. The published WNV genome from Brazil (MH643887) was used to generate a mean 98.4% consensus sequences that formed the target for primer design. The thermocycling conditions involved 40 cycles, and reaction conditions were previously reported in [37].

### 4.3. Library Preparation and Nanopore Sequencing

Amplicons were purified using 1× AMPure XP Beads, and cleaned-up PCR products concentrations were measured using Qubit™ dsDNA HS Assay Kit on a Qubit 3.0 fluorimeter (Thermo Fisher Scientific, Waltham, MA, USA). DNA library preparation was carried out using the Ligation Sequencing Kit and the Native Barcoding Kit (NBD104, Oxford Nanopore Technologies, Oxford, UK) [37]. Purified PCR products pools were pooled together before barcoding reactions (taking in consideration each amplicon pool DNA concentrations), and one barcode was used per sample in order to maximize the number of samples per flow cell. Sequencing library was loaded onto a R9.4 flow cell, and data were collected for up to 6 h, but generally less.

### 4.4. Generation of Consensus Sequences

Raw files were basecalled using Guppy and barcode demultiplexing was performed using qcat. Consensus sequences were generated by de novo assembling using Genome Detective (https://www.genomedetective.com/app/ accessed on 31 December 2019) [38]. New genomes were deposited in the GenBank with accession numbers MW420987, MW420988, and MW420989 (Table 1).

### 4.5. West Nile Virus Typing Tool: Classification Method and Implementation

The classification pipeline we present comprises two components. One for species and sub-species assignment that enables assignment at these levels by BLASTing the query sequences against a set reference sequences [39]. An assignment is made when BLAST reports a result that exceeds the present threshold.

The other component constructs a Neighbor Joining (NJ) phylogenetic tree that is used to make assignments at the lineages and sublineages level. For this component, the query sequence is aligned against a set of reference sequences using the profile alignment option in the ClustalW software [40], such that the query sequence is added to the existing alignment of reference sequences. Following the alignment, a NJ phylogenetic tree with 100 bootstrap replicates is inferred. The tree is constructed using the HKY distance metric with gamma among-site rate variation, as implemented in the PAUP* software (https://paup.phylosolutions.com/ accessed on 31 December 2019) [41]. The query sequence is assigned to a particular genotype if it clusters monophyletically with that genotype clade with bootstrap support >70%. If the bootstrap support is <70%, the genotype is reported to be unassigned (Supplementary Figure S1).

For each of these steps, the earlier discussed reference strains were used with respect to the appropriate typing level (i.e., virus species, lineages, and sublineages). Testing revealed that a BLAST cut-off value of 200 allowed accurate identification of the virus species and WNV using sequence segments >200 base pairs. Note that the species classification procedure is implemented as separate BLAST steps. This enables the tool to efficiently perform large throughput species classification, such as for the classification of shorts sequencing reads. An instance of the web application is publically available on a dedicated server (https://www.genomedetective.com/app/typingtool/wnv/ accessed on 31 December 2019). The web interface on this server accepts up to 2000 whole-genome or partial genome sequences at a time.

### 4.6. Phylogenetic Analysis

The 3 new sequences reported in this study were initially submitted to a genotyping analysis using the new phylogenetic West Nile virus subtyping tool, which is available at https://www.genomedetective.com/app/typingtool/wnv (accessed on 31 December 2019). To put the newly WNV sequences in a global context, we constructed phylogenetic trees to explore the relationship of the sequenced genomes to those of other isolates.

We retrieved 2321 WNV genome sequences with associated lineage date and country of collection from GenBank (Supplementary Figure S2). From this dataset, we generated a subset that included the highly supported (>0.9) clade containing the newly WNV strains obtained in this study plus 29 globally sequences (randomly sampled) from all lineages 1A, 1B, 2, 3, 4, 5, 7, and 8 (Supplementary Table S3). Sequences were aligned using MAFFT [42] and edited using AliView [43]. Those datasets were assessed for the presence of phylogenetic signal by applying the likelihood mapping analysis implemented in the IQ-TREE 1.6.8 software [44]. A maximum likelihood phylogeny was reconstructed using IQ-TREE 1.6.8 software under the HKY+G4 substitution model [44]. We inferred time-scaled trees by using TreeTime [45].

### *4.7. WNV Epidemiological Data*

Human reported cases presenting neurological disease compatible with WNV infection collected between November 2015 and early 2020 were obtained from SINAN. We reinforce the nature of the reports as suspected (not confirmed), being officially defined as cases presenting neurological syndromes compatible with WNV infection, registered as suspected occurences of West Nile virus infection (code A923). As such, the spatio-temporal series of suspected cases should only be interpreted as a proxy for the possible spatio-temporal dynamics of WNV infections [46].

### *4.8. Modeling Transmission Potential*

To estimate the transmission potential of WNV, we employed the computational approach from Lourenço et al. recently applied in Israel [25] and Portugal [26]. This approach estimates the suitability index P using climatic variables only. The index measures the transmission potential of single adult female mosquitoes (spp. Culex) in the animal reservoir and is thus interpreted as a summary measure of the risk for spill-over into human populations. The theory and practice of estimating the index P for mosquito-borne viruses has been previously described in full by Obolski et al. [27]. The epidemiological priors used were the same as in the original study by Lourenço et al. in Israel, which relate to spp. Culex, WNV, and an average bird species. Climatic data were obtained from Copernicus.eu (https://www.copernicus.eu (accessed on 31 December 2019)); in particular, we used the dataset "essential climate variables for assessment of climate variability from 1979 to present" [47]. This dataset offers climatic variables at a time resolution of 1 month and gridded spatial resolution of $0.25 \times 0.25$.

**Supplementary Materials:** The following are available online at https://www.mdpi.com/article/10.3390/pathogens10070896/s1, Figure S1: WNV typing tool, Figure S2: Maximum likelihood phylogenetic tree of 2321 WNV complete genomes. Colors indicates different lineages. Highlighted red clade include the WNV viral strain obtained in this study, Table S1: Primer scheme, Table S2: WNV suspected cases reported between 2014–2020 in each Brazilian state, according to SINAN, Table S3: Globally reference WNV sequences from the subset $n = 29$ used in this study.

**Author Contributions:** Conception and design: É.A.C., M.G., J.L. and L.C.J.A.; Data collection: É.A.C., M.G., L.S.C., V.F., M.A.d.C.e.S.V., D.F.H., D.B.d.A.M., K.R.L.J.C., N.F.O.d.M., A.P.M.R. and L.C.J.A.; Investigations: F.F.A., F.L.L.C., A.M.B.d.F., R.V.d.C., É.A.C., M.G., J.X., V.F., M.I.M.C.G., B.S.Á.d.S.S., A.S.G.S., R.d.P.A.M., N.R.d.C.F., R.F.d.S., R.B.C.T. and J.L.; Data Analysis: M.G., V.F., F.F.A. and J.L.; Writing—Original: É.A.C., M.G., L.S.C., V.F., M.A.d.C.e.S.V., J.L. and L.C.J.A.; Draft Preparation: É.A.C., M.G., L.S.C., V.F., M.A.d.C.e.S.V., J.L. and L.C.J.A.; Revision: É.A.C., M.G., L.S.C., V.F., F.F.A., F.C.d.M.I., M.A.d.C.e.S.V., D.F.H., D.B.d.A.M., M.I.M.C.G., B.S.Á.d.S.S., A.S.G.S., T.d.O., K.R.L.J.C., N.F.O.d.M., A.P.M.R., C.F.C.d.A., L.C.S.F., J.J.M.B., R.B.C.T., O.L.L., S.d.C.S., R.d.P.A.M., R.F.d.S., J.L. and L.C.J.A. Methodology: F.F.A., F.L.L.C. Writing—review & editing: F.F.A., F.L.L.C., A.M.B.d.F., R.V.d.C. All authors have read and agreed to the published version of the manuscript.

**Funding:** This work was founded by CNPq (440685/2016-8, 421598/2018-2), by CAPES (88887.130716/2016-00), by the Pan American Health Organization (IOC-007-FEX-19-2-2-30), by the Fundacão Carlos Chagas Filho de Amparo à Pesquisa do Estado do Rio de Janeiro (FAPERJ, grant number E-26/2002.930/2016 by the European Union's Horizon 2020 Research and Innovation Programme under ZIKAlliance Grant Agreement no. 734548, by the Horizon 2020 through ZikaPlan and ZikAction (grant agreement numbers 734584 and 734857) and by the National Institutes of Health USA grant U01 AI151698 for the United World Antiviral Research Network (UWARN). MG and LCJA is supported by Fundação de Amparo à Pesquisa do Estado do Rio de Janeiro (FAPERJ). JL is supported by a lectureship from the Department of Zoology, University of Oxford.

**Institutional Review Board Statement:** This project was reviewed and approved by the Comissão Nacional de Ética em Pesquisa (CONEP) [National Research Ethics Committee] from the Brazilian Ministry of Health (BrMoH), as part of the arboviral genomic surveillance efforts within the terms of Resolution 510/2016 of CONEP, by the Pan American Health Organization Ethics Review Committee (PAHOERC) (Ref. No. PAHO-2016-08-0029), by the Animal Welfare Committee of Universidade Federal do Piauí, under n°065/19 and by the Oswaldo Cruz Foundation Ethics Committee (CAAE: 90249218.6.1001.5248). All experiments were performed in accordance with relevant guidelines and regulations.

**Informed Consent Statement:** Not applicable.

**Data Availability Statement:** Newly generated WNV sequences have been deposited in GenBank under accession numbers MW420987, MW420988 and MW420989.

**Acknowledgments:** The authors thank the important contributions of the Municipal and Piaui State Health Department (SESAPI, FMS), Municipal and Piaui State Animal Health Department (ADAPI), Laboratório de Saúde Pública do Piauí (LACEN-PI), and the colleague Thiago dos Santos Silva. We also thank the sponsoring institutions: Saint Louis Zoo WildCare Institute and Institute for Conservation Medicine (USA), Universidade Federal do Piauí (UFPI), Fundação de Amparo a Pesquisa do Estado do Piauí (FAPEPI). The authors also thank the Municipal and State Health Department of São Paulo and Minas Gerais state.

**Conflicts of Interest:** The authors declare no conflict of interest.

## References

1. Fall, G.; Di Paola, N.; Faye, M.; Dia, M.; de Melo Freire, C.C.; Loucoubar, C.; de Andrade Zanotto, P.M.; Faye, O. Biological and phylogenetic characteristics of West African lineages of West Nile virus (DWC Beasley, Ed.). *PLoS Negl. Trop. Dis.* **2017**, *11*, 1–23. [CrossRef] [PubMed]
2. Smithburn, K.C.; Hughes, T.P.; Burke, A.W.; Paul, J.H. A neutrotropic virus isolated from the blood of a native of Uganda. *Am. J. Trop. Med. Hyg.* **1940**, *20*, 471–472. [CrossRef]
3. Murgue, B.; Zeller, H.; Deubel, V. The ecology and epidemiology of West Nile virus in Africa, Europe and Asia. *Curr. Top. Microbiol. Immunol.* **2002**, *267*, 195–221. [PubMed]
4. Campbell, G.L.; Marfin, A.A.; Lanciotti, R.S.; Gubler, D.J. West Nile virus. *Lancet Infect. Dis.* **2002**, *2*, 519–529. [CrossRef]
5. Gamino, V.; Höfle, U. Pathology and tissue tropism of natural West Nile virus infection in birds: A review. *Vet. Res.* **2013**, *44*, 46–89. [CrossRef]
6. Bunning, M.L.; Bowen, R.A.; Cropp, B.C.; Sullivan, K.G.; Davis, B.S.; Komar, N.; Godsey, M.; Baker, D.; Hettler, D.L.; Holmes, D.A.; et al. Experimental infection of horses with West Nile virus. *Emerg. Infect. Dis.* **2002**, *8*, 380–386. [CrossRef]
7. Hayes, E.B.; Sejvar, J.J.; Zaki, S.R.; Lanciotti, R.S.; Bode, A.V.; Campbell, G.L. Virology, pathology, and clinical manifestations of West Nile virus disease. *Emerg. Infect. Dis.* **2005**, *11*, 1174–1179. [CrossRef]
8. Kramer, L.D.; Li, J.; Shi, P.Y. West Nile virus. *Lancet Neurol.* **2007**, *6*, 171–181. [CrossRef]
9. Morales, M.A.; Barrandeguy, M.; Fabbri, C.; Garcia, J.B.; Vissani, A.; Trono, K.; Gutierrez, G.; Pigretti, S.; Menchaca, H.; Garrido, N.; et al. West Nile virus isolation from equines in Argentina. *Emerg. Infect. Dis.* **2006**, *12*, 1559–1561. [CrossRef] [PubMed]
10. Osorio, J.E.; Ciuoderis, K.A.; Lopera, J.G.; Piedrahita, L.D.; Murphy, D.; LeVasseur, J.; Carrillo, L.; Ocampo, M.C.; Hofmeister, E. Characterization of West Nile viruses isolated from captive American flamingoes (Phoenicopterus ruber) in Medellin, Colombia. *Am. J. Trop. Med. Hyg.* **2012**, *87*, 565–572. [CrossRef]
11. Martins, L.C.; Silva, E.V.; Casseb, L.M.; Silva, S.P.; Cruz, A.C.; Pantoja, J.A.; Medeiros, D.B.; Martins Filho, A.J.; Cruz, E.D.; Araújo, M.T.; et al. First isolation of West Nile virus in Brazil. *Mem. Inst. Oswaldo Cruz* **2019**, *17*, 114–180332. [CrossRef]
12. Pauvolid-Corrêa, A.; Morales, M.A.; Levis, S.; Figueiredo, L.T.; Couto-Lima, D.; Campos, Z.; Nogueira, M.F.; Silva, E.E.; Nogueira, R.M.; Schatzmayr, H.G. Neutralising antibodies for West Nile virus in horses from Brazilian Pantanal. *Mem. Inst. Oswaldo Cruz* **2011**, *106*, 467–474. [CrossRef]
13. Vieira, M.A.; Romano, A.P.; Borba, A.S.; Silva, E.V.; Chiang, J.O.; Eulálio, K.D.; Azevedo, R.S.; Rodrigues, S.G.; Almeida-Neto, W.S.; Vasconcelos, P.F. West Nile Virus Encephalitis: The First Human Case Recorded in Brazil. *Am. J. Trop. Med. Hyg.* **2015**, *93*, 377–379. [CrossRef]
14. Pauvolid-Corrêa, A.; Campos, Z.; Juliano, R.; Velez, J.; Nogueira, R.M.; Komar, N. Serological evidence of widespread circulation of West Nile virus and other flaviviruses in equines of the Pantanal, Brazil. *PLoS Negl. Trop. Dis.* **2014**, *8*, e2706. [CrossRef] [PubMed]
15. Morel, A.P.; Webster, A.; Zitelli, L.C.; Umeno, K.; Souza, U.A.; Prusch, F.; Anicet, M.; Marsicano, G.; Bandarra, P.; Trainini, G.; et al. Serosurvey of West Nile virus (WNV) in free-ranging raptors from Brazil. *Braz. J. Microbiol.* **2021**, *52*, 411–418. [CrossRef]

16. Melandri, V.; Guimarães, A.É.; Komar, N.; Nogueira, M.L.; Mondini, A.; Fernandez-Sesma, A.; Alencar, J.; Bosch, I. Serological detection of West Nile virus in horses and chicken from Pantanal, Brazil. *Mem. Inst. Oswaldo Cruz* **2012**, *107*, 1073–1075. [CrossRef] [PubMed]
17. Ometto, T.; Durigon, E.L.; de Araujo, J.; Aprelon, R.; de Aguiar, D.M.; Cavalcante, G.T.; Melo, R.M.; Levi, J.E.; de Azevedo Júnior, S.M.; Petry, M.V.; et al. West Nile virus surveillance, Brazil, 2008–2010. *Trans. R. Soc. Trop. Med. Hyg.* **2013**, *107*, 723–730. [CrossRef]
18. Shocket, M.S.; Verwillow, A.B.; Numazu, M.G.; Slamani, H.; Cohen, J.M.; El Moustaid, F.; Rohr, J.; Johnson, L.R.; Mordecai, E.A. Transmission of West Nile and five other temperate mosquito-borne viruses peaks at temperatures between 23 °C and 26 °C. *Elife* **2020**, *9*, e58511. [CrossRef] [PubMed]
19. Haussig, J.M.; Young, J.J.; Gossner, C.M.; Mezei, E.; Bella, A.; Sirbu, A.; Pervanidou, D.; Drakulovic, M.B.; Sudre, B. Early start of the West Nile fever transmission season 2018 in Europe. *Eurosurveillance* **2018**, *23*. [CrossRef] [PubMed]
20. Riccardo, F.; Bolici, F.; Fafangel, M.; Jovanovic, V.; Socan, M.; Klepac, P.; Plavsa, D.; Vasic, M.; Bella, A.; Diana, G.; et al. West Nile virus in Europe: After action reviews of preparedness and response to the 2018 transmission season in Italy, Slovenia, Serbia and Greece. *Global. Health* **2020**, *16*, 47. [CrossRef]
21. Bakonyi, T.; Haussig, J.M. West Nile virus keeps on moving up in Europe. *Eurosurveillance* **2020**, *25*. [CrossRef] [PubMed]
22. Pfeffer, M.; Proebster, B.; Kinney, R.M.; Kaaden, O.R. Genus-specific detection of alphaviruses by a semi-nested reverse transcription-polymerase chain reaction. *Am. J. Trop. Med. Hyg.* **1997**, *57*, 709–718. [CrossRef] [PubMed]
23. Castro-Jorge, L.A.; Siconelli, M.J.L.; Ribeiro, B.D.S.; Moraes, F.M.; Moraes, J.B.; Agostinho, M.R.; Klein, T.M.; Floriano, V.G.; Fonseca, B.A.L.D. West Nile virus infections are here! Are we prepared to face another flavivirus epidemic? *Rev. Soc. Bras. Med. Trop.* **2019**, *52*, e20190089. [CrossRef] [PubMed]
24. Herna´ndez-Triana, L.M.; Jeffries, C.L.; Mansfield, K.L.; Carnell, G.; Fooks, A.R.; Johnson, N. Emergence of West Nile virus lineage 2 in Europe: A review on the introduction and spread of a mosquito-borne disease. *Front. Public Health* **2014**, *2*, 271. [CrossRef] [PubMed]
25. Lourenço, J.; Thompson, R.N.; Thézé, J.; Obolski, U. Characterising West Nile virus epidemiology in Israel using a transmission suitability index. *Eurosurveillance* **2020**, *2*, 5–41.
26. Lourenco, J.; Barros, S.C.; Ze-Ze, L.; Damineli, D.S.; Giovanetti, M.; Osorio, H.C.; Amaro, F.; Henriques, A.M.; Ramos, F.; Luis, T.; et al. West Nile virus in Portugal. *MedRxiv* **2021**. [CrossRef]
27. Obolski, U.; Perez, P.N.; Villabona-Arenas, C.J.; Thézé, J.; Faria, N.R.; Lourenço, J. MVSE: An R-package that estimates a climate-driven mosquito-borne viral suitability index. *Methods Ecol. Evol.* **2019**, *10*, 1357–1370. [CrossRef]
28. Hadfield, J.; Brito, A.F.; Swetnam, D.M.; Vogels, C.B.F.; Tokarz, R.E.; Andersen, K.G.; Smith, R.C.; Bedford, T.; Grubaugh, N.D. Twenty years of West Nile virus spread and evolution in the Americas visualized by Nextstrain. *PLoS Pathog.* **2019**, *15*, e1008042. [CrossRef]
29. Viana, D.S.; Santamarı´a, L.; Figuerola, J. Migratory birds as global dispersal vectors. *Trends Ecol. Evol.* **2016**, *31*, 763–775. [CrossRef]
30. Siconelli, M.J.L.; Jorge, D.M.M.; Castro-Jorge, L.A.; Fonseca-Júnior, A.A.; Nascimento, M.L.; Floriano, V.G.; Souza, F.R.; Queiroz-Júnior, E.M.; Camargos, M.F.; Costa, E.D.L.; et al. Evidence for current circulation of an ancient West Nile virus strain (NY99) in Brazil. *Rev. Soc. Bras. Med. Trop.* **2021**, *54*, e0687–e2020. [CrossRef]
31. Petrone, M.E.; Earnest, R.; Lourenço, J.; Kraemer, M.U.G.; Paulino-Ramirez, R.; Grubaugh, N.D.; Tapia, L. Asynchronicity of endemic and emerging mosquito-borne disease outbreaks in the Dominican Republic. *Nat. Commun.* **2021**, *12*, 151. [CrossRef]
32. Fulop, L.; Barrett, A.D.; Phillpotts, R.; Martin, K.; Leslie, D.; Titball, R.W. Rapid identification of flaviviruses based on conserved NS5 gene sequences. *J. Virol. Methods* **1993**, *44*, 179–188. [CrossRef]
33. Silva, A.S.G.; Matos, A.C.D.; da Cunha, M.A.C.R.; Rehfeld, I.S.; Galinari, G.C.F.; Marcelino, S.A.C.; Saraiva, L.H.G.; Martins, N.R.D.S.; Maranhão, R.P.A.; Lobato, Z.I.P.; et al. West Nile virus associated with equid encephalitis in Brazil, 2018. *Transbound Emerg Dis.* **2019**, *66*, 445–453. [CrossRef] [PubMed]
34. Costa, E.A.; Rosa, R.; Oliveira, T.S.; Assis, A.C.; Paixão, T.A.; Santos, R.L. Molecular characterization of neuropathogenic Equine Herpesvirus 1 Brazilian isolates. *Arq. Bras. Med. Vet. Zootec.* **2015**, *67*, 1183–1187. [CrossRef]
35. Sorg, I.; Metzler, A. Detection of Borna Disease Virus RNA in Formalin-Fixed, Paraffin-Embedded Brain Tissues by Nested PCR. *J. Clin. Microbiol.* **1995**, *4*, 821–823. [CrossRef]
36. Faria, N.R.; Quick, J.; Claro, I.M.; Theze, J.; de Jesus, J.G.; Giovanetti, M.; Kraemer, M.U.; Hill, S.C.; Black, A.; da Costa, A.C.; et al. Establishment and cryptic transmission of Zika virus in Brazil and the Americas. *Nature* **2017**, *546*, 406–410. [CrossRef]
37. Quick, J.; Grubaugh, N.D.; Pullan, S.T.; Claro, I.M.; Smith, A.D.; Gangavarapu, K.; Oliveira, G.; Robles-Sikisaka, R.; Rogers, T.F.; Beutler, N.A.; et al. Multiplex PCR method for MinION and Illumina sequencing of Zika and other virus genomes directly from clinical samples. *Nat. Protoc.* **2017**, *12*, 1261. [CrossRef]
38. Vilsker, M.; Moosa, Y.; Nooij, S.; Fonseca, V.; Ghysens, Y.; Dumon, K.; Pauwels, R.; Alcantara, L.C.; Vanden Eynden, E.; Vandamme, A.M.; et al. Genome Detective: An automated system for virus identification from high-throughput sequencing data. *Bioinformatics* **2019**, *35*, 871–873. [CrossRef]
39. Altschul, S.F.; Gish, W.; Miller, W.; Myers, E.W.; Lipman, D.J. Basic local alignment search tool. *J. Mol. Biol.* **1990**, *21*, 403–410. [CrossRef]

40. Larkin, M.A.; Blackshields, G.; Brown, N.P.; Chenna, R.; McGettigan, P.A.; McWilliam, H.; Valentin, F.; Wallace, I.M.; Wilm, A.; Lopez, R.; et al. Clustal W and Clustal X version 2.0. *Bioinformatics* **2007**, *23*, 2947–2948. [CrossRef] [PubMed]
41. Lemey, P.; Salemi, M.; Vandamme, A. *The Phylogenetic Handbook: A Practical Approach to Phylogenetic Analysis and Hypothesis Testing*, 2nd ed.; Cambridge University Press: Cambridge, UK, 2009.
42. Katoh, K.; Kuma, K.I.; Toh, H.; Miyata, T. MAFFT version 5: Improvement in accuracy of multiple sequence alignment. *Nucleic Acids Res.* **2005**, *33*, 511–518. [CrossRef] [PubMed]
43. Larsson, A. AliView: A fast and lightweight alignment viewer and editor for large data sets. *Bioinformatics* **2014**, *30*, 3276–3278. [CrossRef] [PubMed]
44. Nguyen, L.T.; Schmidt, H.A.; Von Haeseler, A.; Minh, B.Q. IQ-TREE: A fast and effective stochastic algorithm for estimating maximum-likelihood phylogenies. *Mol. Biol. Evol.* **2015**, *32*, 268–274. [CrossRef] [PubMed]
45. Sagulenko, P.; Puller, V.; Neher, R.A. TreeTime:Maximum-likelihood phylodynamic analysis. *Virus Evol.* **2018**, *4*, vex042. [CrossRef] [PubMed]
46. Ministério da Saúde (MS). Secretaria de Vigilância em Saúde. In *Monitoramento da Febre do Nilo Ocidental no Brasil, 2014 a 2019 (Nota Informativa)*; MS: Brasília, Brazil, 2019; 7p. Available online: https://antigo.saude.gov.br/images/pdf/2019/julho/08/informe-febre-niloocidental-n1-8jul19b.pdf (accessed on 31 December 2019).
47. Copernicus Climate Data Store. Available online: https://cds.climate.copernicus.eu/cdsapp#!/dataset/ecv-for-climate-change?tab=overview (accessed on 23 December 2020).

*pathogens*

MDPI

*Article*

# Pathogenesis of Two Western Mediterranean West Nile Virus Lineage 1 Isolates in Experimentally Infected Red-Legged Partridges (*Alectoris rufa*)

**Virginia Gamino [1], Elisa Pérez-Ramírez [2], Ana Valeria Gutiérrez-Guzmán [1], Elena Sotelo [2], Francisco Llorente [2], Miguel Ángel Jiménez-Clavero [2,3] and Ursula Höfle [1,*]**

1 Grupo SaBio (Sanidad y Biotecnología), Instituto de Investigación en Recursos Cinegéticos (IREC) (CSIC-UCLM-JCCM), 13071 Ciudad Real, Spain; gamino.virginia@gmail.com (V.G.); valerivet@hotmail.com (A.V.G.-G.)

2 Centro de Investigación en Sanidad Animal (CISA) del Instituto Nacional de Investigación y Tecnología Agraria y Alimentaria (INIA-CSIC), 28130 Madrid, Spain; elisaperezramirez@gmail.com (E.P.-R.); e.sotelo@zendal.com (E.S.); dgracia@inia.es (F.L.); majimenez@inia.es (M.Á.J.-C.)

3 Centro de Investigación Biomédica en Red de Epidemiologia y Salud Pública (CIBERESP), 28029 Madrid, Spain

* Correspondence: ursula.hofle@uclm.es

**Citation:** Gamino, V.; Pérez-Ramírez, E.; Gutiérrez-Guzmán, A.V.; Sotelo, E.; Llorente, F.; Jiménez-Clavero, M.Á.; Höfle, U. Pathogenesis of Two Western Mediterranean West Nile Virus Lineage 1 Isolates in Experimentally Infected Red-Legged Partridges (*Alectoris rufa*). *Pathogens* **2021**, *10*, 748. https://doi.org/10.3390/pathogens10060748

Academic Editor: El-Sayed Mohammed Abdel-Whab

Received: 8 May 2021
Accepted: 10 June 2021
Published: 13 June 2021



**Abstract:** West Nile virus (WNV) is the most widespread flavivirus in the world with a wide vertebrate host range. Its geographic expansion and activity continue to increase with important human and equine outbreaks and local bird mortality. In a previous experiment, we demonstrated the susceptibility of 7-week-old red-legged partridges (*Alectoris rufa*) to Mediterranean WNV isolates Morocco/2003 and Spain/2007, which varied in virulence for this gallinaceous species. Here we study the pathogenesis of the infection with these two strains to explain the different course of infection and mortality. Day six post-inoculation was critical in the course of infection, with the highest viral load in tissues, the most widespread virus antigen, and more severe lesions. The most affected organs were the heart, liver, and spleen. Comparing infections with Morocco/2003 and Spain/2007, differences were observed in the viral load, virus antigen distribution, and lesion nature and severity. A more acute and marked inflammatory reaction (characterized by participation of microglia and CD3+ T cells) as well as neuronal necrosis in the brain were observed in partridges infected with Morocco/2003 as compared to those infected with Spain/2007. This suggests a higher neurovirulence of Morocco/2003, probably related to one or more specific molecular determinants of virulence different from Spain/2007.

**Keywords:** West Nile virus lineage 1; pathogenesis; neurovirulence; red-legged partridge; antigen distribution; inflammatory reaction

## 1. Introduction

West Nile virus (WNV) is an arthropod-borne flavivirus whose natural cycle involves bird hosts and mosquito vectors, with horses and humans as accidental or dead-end hosts [1]. Currently, it is considered one of the most widely distributed arboviruses in the world, causing numerous human, equine, and bird outbreaks and mortalities, both in the Old and New World [2,3]. In the Mediterranean basin, WNV activity is continuously increasing, and has been associated with several outbreaks affecting mainly humans and horses [4,5]. In this area, avian mortality due to WNV lineage 1 (L1) has been sporadic, while lineage 2 (L2) WNV has been responsible for significant outbreaks in central Europe and in magpies in Italy and Greece [6–13]. Nevertheless, under laboratory conditions in experimental infections, at least some Mediterranean L1 WNV strains have proven to be pathogenic for European wild bird species [14–18].

In an experimental study we demonstrated that the red-legged partridge (*Alectoris rufa*), a Mediterranean endemic gallinaceous bird species [19], is susceptible to the infection with two lineage 1 Western Mediterranean WNV strains, Morocco/2003 [20] and Spain/2007 [17]. However, the virulence of both strains differed, with 70% mortality in Morocco/2003-infected partridges as compared to 30% mortality in Spain/2007-infected birds, corroborating the results observed in a previous study in a mouse model [21].

The objective of the present work was to study the pathogenesis of WNV in the partridges in the mentioned experimental infection in detail to explain the different infection course and mortality observed between the two WNV strains. For this purpose, we compared the viral load, dynamics of virus appearance, and distribution and severity of microscopic lesions in different tissues. Additionally, we studied the dynamics of inflammatory cell activation and recruitment into the central nervous system (CNS) of the infected partridges. The objective was twofold: on one hand, to determine the evolution of the immune response in the CNS of the avian host, as this information is scarce for birds; and on the other hand, to determine differences in the response induced by the two WNV strains, as this could be considered a virulence marker [22].

## 2. Results

### 2.1. Clinical Signs and Gross Pathology

As described in our previous study [17], clinical signs included loss of appetite, ruffled feathers, paralysis, and lack of responsiveness. These started earlier (4 vs. 5 dpi) and were more severe in partridges infected with Morocco/2003. Macroscopic lesions were observed in all euthanized animals from 3 dpi on. The most affected organs were the heart, spleen, liver, and kidney. While the heart and kidney showed lesions from 3 dpi, the spleen and liver were not affected until 6 dpi. Main macroscopic lesions were pallor of the myocardium and hepatic, splenic, and renal parenchyma, as well as the presence of diffuse petechiae. At 14 dpi, congestion of the kidney and lung were also observed.

### 2.2. Virus Genome Detection

All tissue samples collected from euthanized animals at 3 and 6 dpi tested positive by RRT-PCR. The lowest Ct values (i.e., highest viral loads) were reached at 6 dpi (Table 1). The viral load in tissues, estimated using Ct values, was similar in Morocco/2003- and Spain/2007-infected partridges (or slightly higher for Morocco/2003) (Table 1). At 3 and 6 dpi, the highest viral loads were found in the spleen, kidney, and heart (Table 1). High amounts of viral RNA were also detected in feather pulps collected at 6 dpi (Table 1). At 14 dpi, viral loads were low in most cases, and several tissues were negative in the Spain/2007-infected partridges (Table 1).

### 2.3. Histopathology

Microscopic lesions appeared as early as 3 dpi but were more severe and widespread at 6 dpi (Table 2). At that time, the most affected tissues were the heart, lung, spleen, liver, and bursa of Fabricius (Table 2). The main microscopic finding was the presence of inflammatory infiltrates from 3 dpi onwards. These were mainly composed of lymphocytes, macrophages, and plasma cells. Heterophils were also detected in the liver, spleen, and intestine and, to a lesser extent, in the heart, feather follicles (Figure 1A), and skin of both groups. Cellular degeneration and/or necrosis were especially severe in the heart and liver at 6 dpi and in the kidney at 14 dpi (Figure 1B–C, Table 2). On those days, the liver and spleen showed hemosiderosis. In the brain, endothelial cell swelling was observed throughout the experiment in both groups; however, the development of a non-suppurative encephalitis and neuronal necrosis occurred earlier in partridges infected with Morocco/2003 (Figure 1D, Table 2).

**Table 1.** Viral genome load in tissues of experimentally WNV-infected red-legged partridges. The presence of WNV genome was analyzed in different tissues at different days post-inoculation (dpi) by real-time reverse transcription polymerase chain reaction (RRT-PCR). Data are presented as Ct values. SP07: Spain/2007, MO03: Morocco/2003, NA: tissue sample not analyzed, −: tissue sample with Ct $\geq$ 40.

| | **3 dpi** | | | | **6 dpi** | | | | **14 dpi** | | |
|---|---|---|---|---|---|---|---|---|---|---|---|
| | **SP07** | | **MO03** | | **SP07** | | **MO03** | | **SP07** | | **MO03** |
| Tissue | No. 1 | No. 2 | No. 3 | No. 4 | No. 5 | No. 6 | No. 7 | No. 8 | No. 9 * | No. 10 | No. 11 |
| Brain | 30.3 | 32.8 | 31.4 | 33.9 | 25.2 | 26.5 | 25.2 | 24.8 | NA | − | 37.6 |
| Heart | 28.7 | 28.5 | 25.4 | 24.8 | 19.6 | 20.3 | 23.7 | 18.8 | NA | 38.0 | 33.8 |
| Lung | NA | 30.9 | 31.0 | 30.1 | 26.3 | 25.7 | 29.2 | 24.2 | NA | − | 38.8 |
| Liver | NA | 30.2 | 29.6 | 30.9 | 28.7 | 25.7 | 30.8 | 28.2 | NA | − | − |
| Spleen | 25.3 | 23.7 | 26.5 | 24.4 | 29.7 | 22.7 | 24.6 | 21.8 | NA | 36.1 | 34.0 |
| Kidney | 26.3 | 27.5 | 24.5 | 26.0 | 23.0 | 22.4 | 23.7 | 20.9 | NA | 37.5 | 36.1 |
| Thymus | 27.6 | 29.3 | 31.9 | 23.5 | 30.2 | 23.4 | 33.0 | NA | NA | − | 29.0 |
| Bursa of Fabricius | NA | 32.0 | 32.8 | 34.2 | 27.9 | 24.8 | 31.5 | 29.8 | NA | − | 31.3 |
| Feather pulp | − | 31.4 | 27.4 | 31.3 | 18.2 | 20.8 | 22.8 | 16.9 | NA | 34.6 | 33.7 |

* Tissues from individual No. 9 were not available for analysis by RRT-PCR; however, they were used for histopathology and immunohistochemistry analyses.

**Table 2.** Microscopic lesions in tissues of experimentally WNV-infected red-legged partridges. Lesions were graded according to their distribution and severity at different days post-inoculation (dpi) in both groups, Spain/2007 (SP07) and Morocco/2003 (MO03). −: no lesion, +: focal and mild or moderate/multifocal and mild, ++: focal and marked/multifocal and moderate/diffuse and mild, +++: multifocal and marked/diffuse and moderate or marked. NA: tissue sample not analyzed.

| | **3 dpi** | | | | **6 dpi** | | | | **14 dpi** | | |
|---|---|---|---|---|---|---|---|---|---|---|---|
| | **SP07** | | **MO03** | | **SP07** | | **MO03** | | **SP07** | | **MO03** |
| Tissue/Lesion | No. 1 | No. 2 | No. 3 | No. 4 | No. 5 | No. 6 | No. 7 | No. 8 | No. 9 | No. 10 | No. 11 |
| **Cerebrum** | | | | | | | | | | | |
| Neuronal necrosis | − | − | − | − | − | − | + | + | + | − | + |
| Gliosis | − | − | − | − | − | + | + | ++ | + | + | − |
| Perivascular cuffing | − | − | − | − | − | − | − | − | − | + | + |
| Endothelial cell swelling | + | + | ++ | ++ | + | + | + | ++ | ++ | + | + |
| **Cerebellum** | | | | | | | | | | | |
| Purkinje cell necrosis | − | − | − | − | − | − | ++ | + | + | NA | + |
| Gliosis | − | − | − | − | − | − | ++ | + | − | NA | ++ |
| Perivascular cuffing | − | − | − | − | − | − | − | − | + | NA | − |
| Endothelial cell swelling | − | − | ++ | ++ | + | + | +++ | ++ | ++ | NA | ++ |

**Table 2.** *Cont.*

| | 3 dpi | | | | 6 dpi | | | | 14 dpi | | |
|---|---|---|---|---|---|---|---|---|---|---|---|
| | SP07 | | MO03 | | SP07 | | MO03 | | SP07 | | MO03 |
| **Heart** | | | | | | | | | | | |
| Myofiber necrosis-degeneration | + | + | – | + | +++ | ++ | +++ | +++ | ++ | + | ++ |
| Inflammatory infiltrate | – | ++ | – | ++ | +++ | +++ | +++ | +++ | + | ++ | +++ |
| **Lung** | | | | | | | | | | | |
| Inflammatory infiltrate | – | +++ | – | ++ | +++ | +++ | +++ | +++ | + | – | +++ |
| **Liver** | | | | | | | | | | | |
| Hepatocyte necrosis | – | – | – | – | – | +++ | – | +++ | – | – | – |
| Inflammatory infiltrate | – | ++ | ++ | ++ | ++ | ++ | ++ | +++ | – | ++ | + |
| Hemosiderosis | – | – | – | – | + | + | ++ | ++ | ++ | + | + |
| Tissue/Lesion | No. 1 | No. 2 | No. 3 | No. 4 | No. 5 | No. 6 | No. 7 | No. 8 | No. 9 | No. 10 | No. 11 |
| **Spleen** | | | | | | | | | | | |
| Lymphoid cell necrosis | – | – | – | – | – | – | – | – | ++ | – | – |
| Lymphoid cell depletion | – | – | ++ | + | + | + | + | + | – | – | – |
| Granulocytic infiltrate | +++ | +++ | ++ | ++ | + | +++ | – | – | +++ | ++ | +++ |
| Eosinophilic material deposits | – | – | – | – | +++ | +++ | ++ | ++ | ++ | – | ++ |
| Hemosiderosis | – | – | – | – | +++ | +++ | + | +++ | ++ | – | + |
| **Kidney** | | | | | | | | | | | |
| Tubular epithelial cell necrosis | – | – | – | – | – | – | – | – | +++ | +++ | +++ |
| Inflammatory infiltrate | – | ++ | – | ++ | – | – | – | + | + | – | ++ |
| **Duodenum** | | | | | | | | | | | |
| Inflammatory infiltrate | + | + | – | – | – | + | – | – | NA | – | + |
| **Large intestine** | | | | | | | | | | | |
| Inflammatory infiltrate | – | + | – | ++ | – | – | – | – | NA | – | ++ |
| **Cecal tonsils** | | | | | | | | | | | |
| Lymphoid cell necrosis | NA | NA | NA | + | NA | ++ | NA | NA | NA | – | NA |
| **Bursa of Fabricius** | | | | | | | | | | | |
| Lymphoid cell necrosis | – | – | – | – | NA | – | – | ++ | NA | NA | – |
| Lymphoid cell depletion | – | + | – | – | NA | ++ | ++ | ++ | NA | NA | – |
| **Skin + feather follicle** | | | | | | | | | | | |
| Inflammatory infiltrate skin | – | NA | – | +++ | – | NA | NA | – | NA | ++ | – |
| Inflammatory infiltrate feather pulp | +++ | NA | +++ | NA | NA | NA | NA | +++ | NA | – | +++ |

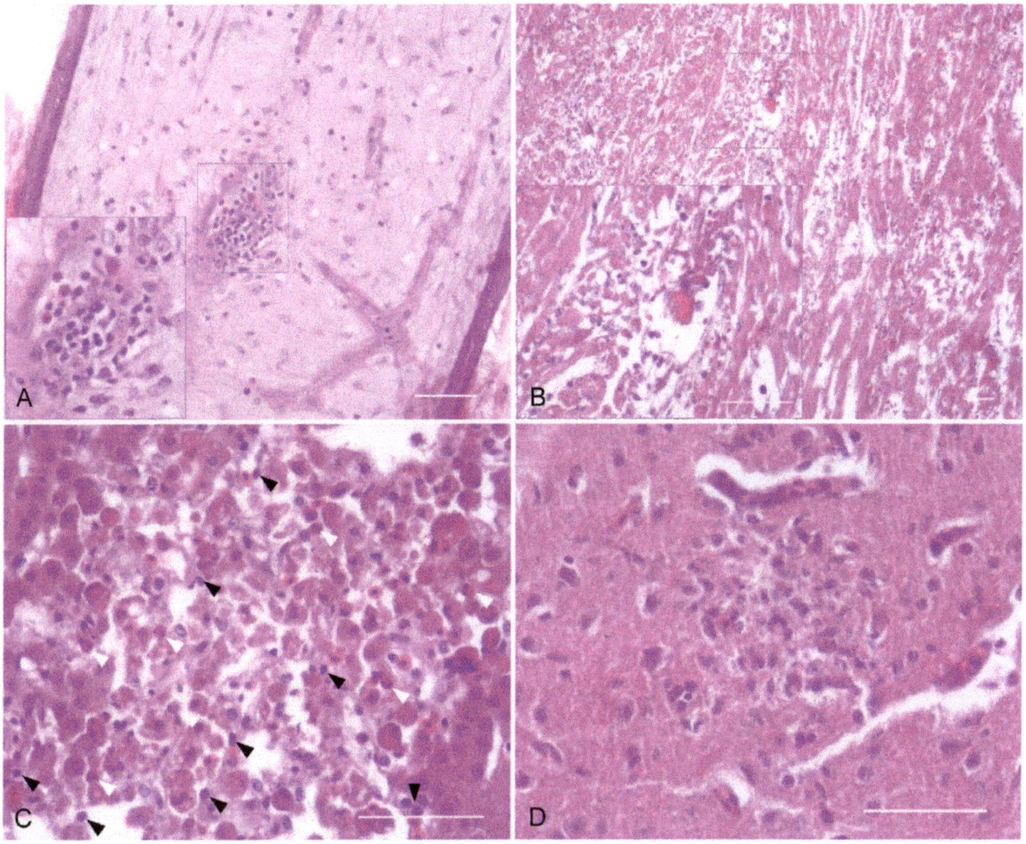


**Figure 1.** Microscopic lesions in tissues of experimentally WNV-infected red-legged partridges. (**A**) Feather follicle; partridge inoculated with Morocco/2003, 14 dpi. Multifocal infiltration of inflammatory cells in the feather pulp. Inset: detail of an inflammatory nodule composed of lymphocytes, macrophages, and granulocytes. (**B**) Heart; partridge inoculated with Spain/2007, 6 dpi. Diffuse and marked necrosis and degeneration of cardiac myofibers, and infiltration of inflammatory cells. Inset: detail of necrotic cardiac myofibers which show degeneration, fragmentation, and accumulation of hyaline material in the cytoplasm. Infiltration of mononuclear inflammatory cells is also observed. (**C**) Liver; partridge inoculated with Morocco/2003, 6 dpi. Necrosis of hepatocytes characterized by cellular detachment, lysis of the cytoplasm and pyknosis and fragmentation of the nucleus. There is also a mild infiltration of lymphocytes and plasma cells (black arrowheads) and heterophils (white arrowheads). (**D**) Cerebrum; partridge inoculated with Morocco/2003, 6 dpi. Focal necrosis with degeneration of the neuropil and infiltration of lymphocytic and glial cells. Scale bars = 100 μm.

As an incidental finding, there were coccidian oocysts and gametes in the enterocytes of the large intestine. These parasites were much more numerous in WNV-infected birds as compared to control individuals (Figure 2A), especially in partridges euthanized at 6 dpi and in those infected with Morocco/2003. In fact, one bird of this group showed an associated severe necrosis of the epithelium of intestinal mucosa and crypts (Figure 2B).

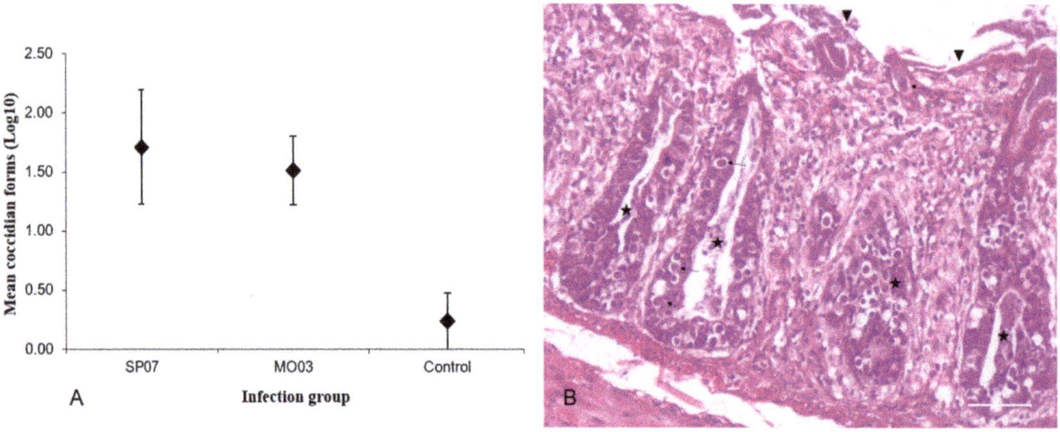


**Figure 2.** Coccidiosis in the large intestine of WNV-infected and control red-legged partridges. (**A**) Graph showing the mean numbers ± standard deviation of log transformed nos. of coccidian forms in transverse sections of the large intestine at the level of cecal tonsils in red-legged partridges experimentally infected with WNV Morocco/2003 (MO03) and Spain/2007 (SP07) strains. (**B**) Cecum; partridge inoculated with Morocco/2003, 6 dpi. Presence of gametes and coccidian oocysts in the epithelium of the mucosa and crypts (arrows). There is also inflammation in the lamina propria and severe necrosis of the epithelium of both the mucosa (arrowheads) and crypts (stars). Scale bar = 100 μm.

### *2.4. Virus Antigen*

Virus antigen staining was mild to moderate in most cases. Immunopositivity was more widespread at 6 dpi and in the tissues of partridges infected with Morocco/2003 (Table 3), consistent with RRT-PCR results. At 3 dpi, WNV antigen was detected in macrophages in spleen and inflammatory cells and myofibers of the heart of both groups. In Morocco/2003-infected partridges there was also a mild immunostaining in tubular epithelial cells of the kidney, isolated acinar cells of the pancreas, and a small group of inflammatory cells in the cecal tonsils. At 6 dpi, the WNV antigen was stained in inflammatory cells in the lung, heart, spleen, kidney, and pancreas. It was also detected in cardiac myofibers (Figure 3A), glomerular mesangial cells (Figure 3B), and tubular epithelial cells of the kidney, acinar cells of the pancreas, as well as in cells of the crypts and myofibers of the muscularis externa of the large intestine. Only in birds infected with Morocco/2003 was the WNV antigen evidenced in smooth muscle cells of the splenic vessels, hepatocytes, and Kupffer cells of the liver, as well as smooth muscle cells of the vascular wall of one vessel and several myofibers of the muscularis externa of the duodenum. At 14 dpi, IHC was negative in all examined tissues (Table 3).

### *2.5. Inflammatory Cells in the Brain*

#### 2.5.1. Microglia Activation/Macrophage Infiltration

At 3 dpi, a mild reaction of microglial cells was observed in Morocco/2003-infected partridges. Nevertheless, it was at 6 dpi when these cells were more active in both groups, especially in birds infected with Morocco/2003 (Figure 4A). At 14 dpi, while the activity of microglial cells decreased in the aforementioned group, these remained very active in Spain/2007-infected individuals (Figure 4A).

**Table 3.** WNV antigen detection in tissues of experimentally WNV-infected red-legged partridges. WNV antigen was detected by immunohistochemistry at different days post-inoculation (dpi) in both groups, Spain/2007 (SP07) and Morocco/2003 (MO03). Immunostaining was graded according to its distribution and percentage of stained cells. –: no staining, ±: focal single cells, +: focal or multifocal and <20% cells stained, ++: multifocal or diffuse and 20–50% cells stained, +++: multifocal or diffuse and >50% cells stained. NA: tissue sample not analyzed.

| | **3 dpi** | | | | **6 dpi** | | | | **14 dpi** | | |
|---|---|---|---|---|---|---|---|---|---|---|---|
| | **SP07** | | **MO03** | | **SP07** | | **MO03** | | **SP07** | | **MO03** |
| Tissue | No. 1 | No. 2 | No. 3 | No. 4 | No. 5 | No. 6 | No. 7 | No. 8 | No. 9 | No. 10 | No. 11 |
| Heart | ± | – | + | + | ++ | ++ | + | +++ | – | – | – |
| Lung | – | – | – | – | – | ± | – | + | – | – | – |
| Liver | – | – | – | – | – | – | – | + | – | – | – |
| Spleen | + | + | + | ± | – | + | + | + | – | – | – |
| Kidney | – | – | + | + | + | + | + | ++ | – | – | – |
| Duodenum | – | – | – | – | – | – | – | + | NA | – | – |
| Large intestine | – | – | – | – | + | + | + | + | NA | – | – |
| Pancreas | – | – | ± | ± | + | + | + | + | NA | – | – |
| Cecal tonsils | NA | NA | NA | + | NA | – | NA | NA | NA | – | NA |

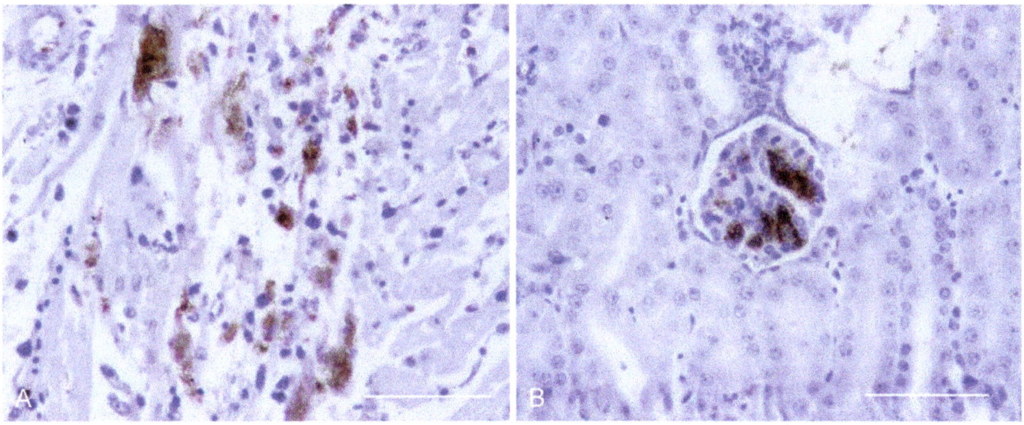


**Figure 3.** Detection of WNV antigen by immunohistochemistry in tissues of experimentally-infected red-legged partridges. (**A**) Heart; partridge inoculated with Spain/2007, 6 dpi. WNV antigen in the cytoplasm of cardiac myofibers. (**B**) Kidney; partridge inoculated with Morocco/2003, 6 dpi. WNV antigen in the cytoplasm of glomerular mesangial cells. Scale bars = 100 μm.

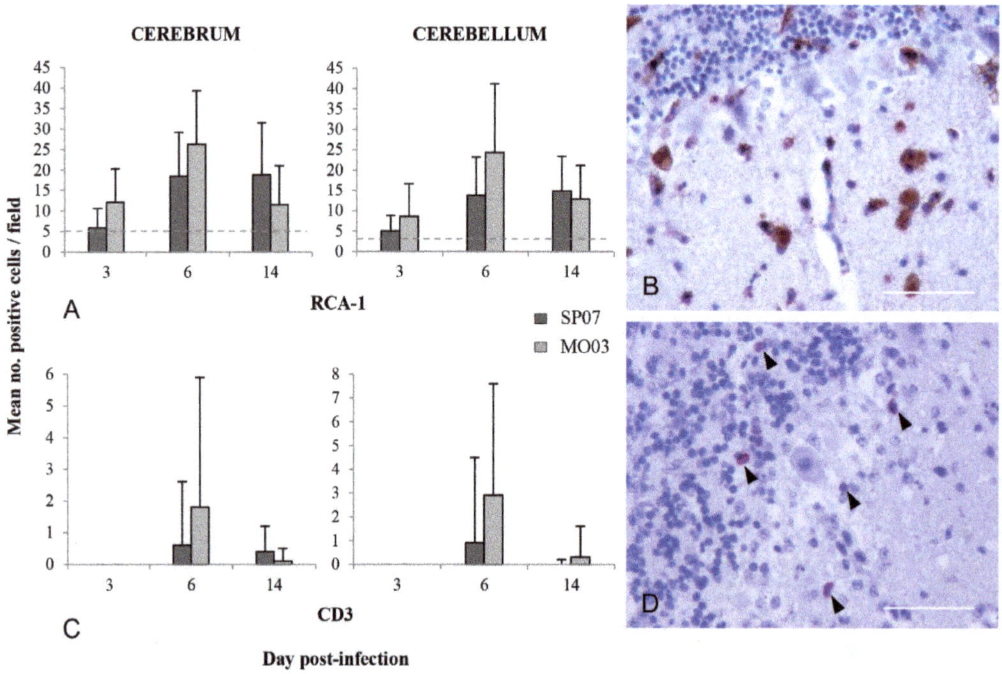


**Figure 4.** Characterization of inflammatory cell reaction in the brain of red-legged partridges experimentally infected with WNV Morocco/2003 (MO03) and Spain/2007 (SP07) strains. (**A**) Semi-quantitative analysis of the number of RCA-1+ microglial cells/macrophages in the cerebrum and cerebellum at different days post-inoculation (dpi). Bars represent the mean of stained cells in 30 randomly selected fields (at 400x magnification) ± standard deviation. The horizontal dashed line represents the mean of stained cells in the cerebrum and cerebellum of a non-WNV-infected partridge. (**B**) Cerebellum; partridge inoculated with Morocco/2003, 6 dpi. RCA-1 positive staining in the cytoplasm of phagocytic foamy macrophages in the molecular layer. Scale bar = 100 µm. (**C**) Semi-quantitative analysis of the number of CD3+ T cells in the cerebrum and cerebellum at different dpi. Bars represent the mean of stained cells in 30 randomly selected fields (at 400x magnification) ± standard deviation. (**D**) Cerebellum; partridge inoculated with Morocco/2003, 6 dpi. CD3 positive staining in T cells located near the Purkinje cell layer (arrowheads). Scale bar = 100 µm.

In the cerebrum, at 3 dpi, RCA-1+ ramified cells were diffusely distributed in the parenchyma. At that time, there was also the mild presence of amoeboid (large soma and short, thick cellular processes) and rounded cells (corresponding both to activated microglia and macrophages). From 6 dpi onwards, most microglia changed to an activated amoeboid morphology and increased the presence of microglia/macrophage nodules, which in many cases surrounded neurons and vessels. RCA-1+ cells were especially abundant in the peripheral pallium and in the region located near the lateral ventricle. In the cerebellum, microglial cell reaction was milder than in the cerebrum (Figure 4A), but the evolution of changes in cellular morphology was similar. These cells were more active in the molecular layer, some of them surrounding Purkinje cells (Figure 4B). At 6 dpi, cell activation was also moderate in the granular layer and white matter, especially in Morocco/2003-infected partridges.

2.5.2. Astrocyte Activation

GFAP+ cells were detected from 3 dpi on, with mild changes in their distribution and relative abundance during the infection course and between infected groups. Compared to a non-infected partridge, there were mild differences in staining distribution and the quantity of stained cells but none in astrocyte morphology or staining intensity.

In the cerebrum, moderate astrocytosis (i.e., increased number of astrocytes) was detected near the lateral ventricle in both groups and in the lamina medularis dorsalis in Morocco/2003-infected birds. In the cerebellum, there was mild astrocytosis in the granular layer, slightly more marked at 6 dpi and in Morocco/2003-infected partridges. In this group, and especially in one bird euthanized at 6 dpi, some GFAP+ fibers invaded the molecular layer and among Purkinje cells. Moderate astrocytosis was also observed in the white matter of Morocco/2003-infected birds.

#### 2.5.3. T Cell Infiltration

CD3+ T cells infiltrated the brain parenchyma after 3 dpi, and at 6 dpi these cells were more numerous, especially in Morocco/2003-infected birds (Figure 4C). At 14 dpi, the number of T cells decreased, more sharply in the Morocco/2003-infected group (Figure 4C). In some cases, brain zones of T cell infiltration corresponded to brain zones of microglia activation and/or macrophage infiltration. Few T cells were detected in meningeal vessels, and only in Morocco/2003-infected partridges.

In the cerebrum, CD3+ T cells were found diffusely distributed, but also forming part of perivascular infiltrates and of inflammatory foci/nodules associated, in some cases, with neuronal necrosis. The peripheral pallium was especially infiltrated by these cells. In the cerebellum, the vast majority of CD3+ T cells were distributed in the molecular layer, in some cases forming part of inflammatory nodules surrounding Purkinje cells (Figure 4D). There was also moderate infiltration in the granular layer and mild infiltration in the white matter.

## 3. Discussion

In this work, we studied differences in the pathogenesis after experimental infection of red-legged partridges with two different Mediterranean L1 WNV isolates, Morocco/2003 and Spain/2007, to elucidate the cause of the different infection course and mortality observed during the experiment and described in a previous work [17].

In both experimental groups, the WNV genome and antigen as well as macroscopic and microscopic lesions were detected as early as 3 dpi. Nevertheless, day 6 post-inoculation (3 days after peak viremia, [17]) can be considered critical in the course of the infection, since at that time, the viral load in most tissues reached the highest levels, the virus antigen was more widespread and abundant, and microscopic lesions were more severe. At 14 dpi, although microscopic lesions were still observed, the low viral loads in organs and the absence of virus antigen detectable by IHC suggest that, by this dpi, the partridges had cleared most of the infecting virus. Nevertheless, potential virus maintenance in some tissues of a low percentage of infected individuals should be considered, as persistent infection has been demonstrated in experimentally and naturally WNV-infected birds [23,24].

Microscopic lesions found in the partridges were consistent with those described in other WNV-infected gallinaceous birds [25–27]. Nevertheless, while we found endothelial cell swelling, gliosis, and neuronal necrosis in the brain, CNS lesions were not described in chukar partridges (*Alectoris chukar*) naturally infected with WNV [26,28]. In contrast, in experimentally infected ruffed grouse (*Bonasa umbellus*) who were native from the North American continent, encephalitis of differing severity was present in all, even vaccinated, individuals [29]. Differences in infection conditions, such as the specific virus strain, inoculation dose, and age or bird species, could account for the differences in pathological findings [18,30]. The marked presence of inflammatory infiltrates in feather pulps and skin from 3 dpi onwards was probably associated with the high viral load found by RRT-PCR. WNV has been detected in the skin and feather follicles in naturally and experimentally infected birds [31–33], and feather-picking has been suggested as a potential way of horizontal WNV transmission [34]. The exacerbation of coccidian oocysts infestation in the intestines, concurrent with the acute phase of WNV infection (3–6 dpi), suggests a potential WNV-mediated effect on immune function, although more studies are necessary to support this hypothesis. Secondary pathological processes in WNV-infected birds have been documented on numerous occasions [7,13,35,36].

In general, the viral loads detected by RRT-PCR correlated well with the virus antigen detected by IHC and microscopic lesions, with some exceptions such as the feather pulp or the brain. Differences in the results between RRT-PCR and IHC can be explained by their different sensitivity (they detect different viral components), specific areas analyzed, or just by differences in sample collection. In the specific case of the brain, negative results in IHC, despite the mild to moderate neuronal necrosis, could also be explained by the induction of cell injury by the local inflammatory response rather than the direct effect of neuronal virus infection, as has been previously suggested [37–39]. However, considering our RRT-PCR results in this tissue, the lower sensitivity of IHC is a more likely explanation. In comparison free-ranging goshawks (*Accipiter gentilis*) infected with lineage 2 WNV predominantly had CNS lesions and abundant WNV antigen detection in foci associated with lesions, in addition to severe myocarditis and lesions in the liver, spleen, and kidney [40]. Additionally, free-living magpies infected with WNV L2 showed severe CNS signs while, in contrast, experimentally infected red-legged partridges, our study species, had only non-specific signs and macroscopic lesions similar as those observed in WNV L1-infected birds, although in both studies no histopathologic descriptions were included [12,41]. This suggests, on the one hand, that bird-virulent WNV L2 strains may be more neurovirulent while, on the other hand, as reported by other authors [18], WNV pathogenesis may be highly dependent on both the infecting strain and the avian host species.

Regarding the inflammatory response in the CNS, microglial cells were the most reactive population in the CNS of the WNV-infected partridges, mainly at 6 dpi (corresponding with the highest viral load), when most of these cells changed to an activated amoeboid morphology. The other resident immune effector of the CNS, astrocytes, appeared to play a limited role. Some authors have indicated that astrocytes can react later than the microglia and that their maximum activity occurs 14 days after CNS injury [42]. Therefore, it is possible that a more marked astrocytosis and/or astrogliosis would be detected later if our experimental study was extended. Microglia and astrocytes are considered essential in the anti-flavivirus response in the CNS of humans and rodent models [30,43–46]. In response to resident cell activation, among other stimuli, CD3+ T cells infiltrated the brain, especially at 6 dpi, and mainly in the cerebral pallium and molecular layer of the cerebellum. Upon antigen presentation, T cells act by directly destroying virus-infected cells and by producing cytokines that increase immune cell recruitment and stimulate other immune effectors [47]. In humans and rodents, it has been demonstrated that these cells are essential for WNV clearance and for recovery from the disease [43,48–50].

For a global interpretation of the WNV-induced encephalitis in the partridges, it is important to highlight that despite the essential defense role of all these cell components for the recovery from WNV infection; it has been indicated that a robust response can also have detrimental effects, contributing to neuron damage [43,51].

Although similar infection dynamics were found in partridges infected with Morocco/2003 and Spain/2007 strains, some differences were noted related to the viral load, virus antigen distribution, and severity of microscopic lesions. The most striking differences were found in the CNS, where encephalitis and neuronal necrosis were more acute and severe in partridges infected with Morocco/2003. The pathogenesis of WNV infection depends mainly on viral and host factors, which determine the level of viral replication and the severity of the infection [30]. As the infected partridges were homogeneous in terms of age, history of rearing, and maintenance, immunologic status should have been very similar between groups. For this reason, the observed differences were most probably due to factors related to the WNV strain. Genetic changes, particularly those leading to amino acid substitutions, can modify the virulence of WNV strains, as well as other phenotypic traits such as host and/or mosquito competence [52–55]. The Morocco/2003 and Spain/2007 WNV strains differ in 13 amino acid positions; therefore, it is possible that one or several of these changes determined/modulated the virulence of both strains for the red-legged partridge [21].

The pathogenesis of the WNV infection in the CNS depends mainly on the capacity of the virus strain to enter the CNS (neuroinvasiveness) and to produce lesions (neurovirulence) [56]. Some authors have indicated that, once a certain viremia level is reached, the ability of a WNV strain to enter the CNS is much more important than its own neurovirulence [57,58]. Although the mean peak viremia titer was higher in the partridges inoculated with Morocco/2003, there were no statistically significant differences between groups [17], and in both, WNV was present in the brain as early as 3 dpi. At 6 dpi, despite a similar viral load in the brain, a more intense inflammatory reaction and more severe microscopic lesions were observed in Morocco/2003-infected partridges. For these reasons, our results point out that differences in intrinsic neurovirulence between strains were more important than their neuroinvasiveness. It seems plausible that a higher neurovirulence of Morocco/2003, associated with a more exacerbated inflammatory reaction, resulted in a more acute and severe pathology in the CNS. Nevertheless, the exact mechanism by which this strain results in being more neurovirulent remains unknown, and the limited number of birds analyzed in this study forces us to be cautious with the interpretation of the results obtained.

In conclusion, the higher virulence of Morocco/2003 for the red-legged partridge is probably related to a more acute and severe encephalitis in the infected birds. Further studies are needed to elucidate the specific genetic markers that determine this higher neurovirulence.

## 4. Materials and Methods

### *4.1. Viruses*

Two different WNV isolates from the Mediterranean basin were used in this study: Morocco/2003 (strain 04.05, GenBank acc. n°: AY701413) [20] and Spain/2007 (strain GE-1b/B, GenBank acc. n°: FJ766331) [21]. Details on the preparation of the inocula are given in Sotelo et al. [17].

### *4.2. Experimental Infection*

The partridges were obtained, maintained, handled, and inoculated as described in Sotelo et al. [17]. Briefly, two groups of 7-week-old red-legged partridges were subcutaneously inoculated in the cervical region with $10^4$ PFU/individual of either WNV Morocco/2003 or Spain/2007 diluted in up to 0.1 mL Dulbecco's Minimum Essential Medium (DMEM) (supplemented with 2 mM L-glutamine, 100 U/mL penicillin, and 100 µg/mL streptomycin). For the specific purpose of the present study, we performed euthanasia by intravenous injection of embutramide (T61®, Intervet–Schering-Plough, Madrid, Spain) on two birds of each group at days 3 and 6 post-inoculation (dpi), and of two birds infected with Spain/2007 and one infected with Morocco/2003 at 14 dpi.

### *4.3. Sample Collection*

Detailed necropsies were performed on the euthanized individuals. Samples of the brain, heart, lung, liver, spleen, kidney, thymus, bursa of Fabricius, and feather pulp were collected into sterile polypropylene tubes filled with 1 mL of Hanks' balanced solution (10% glycerol, 200 U/mL penicillin, 200 µg/mL streptomycin, 100 U/mL polymixin B sulphate, 250 µg/mL gentamicin, and 50 U/mL nystatin) and stored at −70 °C until analysis by real-time reverse transcription polymerase chain reaction (RRT-PCR). In addition, samples of the brain, oral mucosa, thymus, heart, trachea, lung, liver, spleen, kidney, small and large intestine, pancreas, cecal tonsils, bursa of Fabricius, pectoral muscle, and skin with feather follicles were fixed in 10% neutral buffered formalin.

### 4.4. Virus Genome Detection

RNA was extracted from tissue samples after homogenization and tested by RRT-PCR for the presence of the WNV genome as described in Sotelo et al. [17].

### 4.5. Histopathology

Formalin-fixed tissue samples were trimmed, embedded in paraffin, and processed to obtain hematoxylin and eosin-stained sections. These were independently examined by two different investigators (UH and VG) to determine the presence of WNV-associated lesions. When lesions were present, they were graded according to their distribution (focal, multifocal, or diffuse) and severity (mild, moderate, or marked).

### 4.6. Immunohistochemistry

Tissue sections were mounted on Vectabond™ reagent (Vector Laboratories, Inc., Burlingame, CA, USA)-pretreated slides. Immunohistochemical detection of the WNV antigen was performed using a rabbit polyclonal antibody (BioReliance, Product 81–015, Rockville, MD, USA) at a dilution of 1:1000, following the protocol described previously [59]. We also characterized the inflammatory cell population in the cerebrum and cerebellum. For that purpose, we used primary antibodies, reagents, and protocols detailed in Table 4. Endogenous peroxidase activity was inhibited with a peroxidase-blocking reagent (Dako EnVision®+System-HRP (AEC), DakoCytomation, Carpinteria, CA, USA) (CD3, GFAP) or with 3% H202 diluted in methanol (RCA-1), rinses were performed using 0.1% Tris-buffered saline/Tween20 (TBS 0.05 M, pH 7.5), unspecific primary antibody labeling was blocked with 2% albumin from bovine serum (BSA) (Sigma-Aldrich Chemie, Steinheim, Germany) diluted in 0.1% TBS/Tween20, and sections were counterstained with Mayer's hematoxylin.

**Table 4.** Reagents and protocols used to characterize inflammatory cells in the brain of experimentally WNV-infected red-legged partridges.

| Primary Antibody [a] | Cell Population | Pretreatment [b] | Primary Antibody Dilution and Incubation [c] | Secondary Antibody [d] | Detection System [e] |
|---|---|---|---|---|---|
| Lectin RCA-1 biotinylated | Microglia-macrophages | Citrate buffer Microwave heat (22 min) | 1:600, 45 min RT | Goat anti-rabbit IgG | ABC-DAB |
| Polyclonal rabbit anti-GFAP | Astrocytes | Proteinase K (7 min RT) | 1:500, 4 °C ON | Labelled polymer-HRP anti-rabbit | AEC + substrate chromogen |
| Polyclonal rabbit anti-human CD3 | T cells | Citrate buffer Microwave heat (22 min) | 1:500, 4 °C ON | Labelled polymer-HRP anti-rabbit | AEC + substrate chromogen |

[a] Primary antibody products: RCA-1 product No. B-1085 (Vector Laboratories); GFAP product No. Z0334 (DakoCytomation, Glostrup, Denmark); CD3 product No. A0452 (DakoCytomation); CD79a product No. RM-9118 (Thermo Fisher Scientific, Runcorn, UK). [b] Proteinase K (DakoCytomation); RT: room temperature (22–25 °C). [c] Antibodies were diluted in 2% BSA−0.1% TBS/Tween20. ON: overnight. [d] Goat anti-rabbit IgG (Vector Laboratories) was diluted 1:200 in 0.1% TBS/Tween20 and applied for 1 h at RT; Labelled polymer-HRP anti-rabbit (Dako EnVision®+System-HRP (AEC), DakoCytomation) was applied according to manufacturer's recommendation. [e] Avidin-biotinylated enzyme complex (ABC system, Vector Laboratories) was applied for 30 min according to the manufacturer's recommendation and 3,3′-diaminobenzidine tetrahydrochloride (DAB, Vector Laboratories) was applied for 30 s according to the manufacturer's recommendations. AEC+substrate chromogen (Dako EnVision®+System-HRP (AEC), DakoCytomation) was applied for 15 min (CD3, CD79) and 3 min (GFAP) according to the manufacturer's recommendations.

Tissue sections of WNV RRT-PCR-positive red-legged partridges were used as positive controls. Controls of specificity included several sections with substitution of the primary antibody by 2% BSA−0.1% TBS/Tween20 and negative rabbit antibody (BioReliance, Product 81–015), and tissue sections of a non-infected (WNV RRT-PCR-negative) red-legged partridge (from sham-inoculated control group) [17]. For the detection of T cells, the positive control included a section of spleen of non-infected red-legged partridges. Negative controls included substitution of the primary antibody by 2% BSA−0.1% TBS/Tween20 and a brain section of a WNV RRT-PCR-negative partridge. A brain section of a non-WNV-infected partridge of the same age (control group [16]) served as reference for RCA-1 and GFAP.

We scored virus antigen staining according to its distribution and abundance in the tissues. The distribution of inflammatory cells within the brain was evaluated at 200x magnification. To detect changes in the abundance of CD3+ T cells and RCA-1+ cells during the course of infection, we counted the number of stained cells in 30 randomly selected fields at 400x magnification (in each brain region: cerebrum and cerebellum). Changes in the morphology, abundance, and staining intensity of GFAP+ astrocytes were also evaluated.

**Author Contributions:** Conceptualization, M.Á.J.-C. and U.H.; Funding acquisition, U.H.; Investigation, V.G., E.P.-R., E.S., M.Á.J.-C., and U.H.; Methodology, V.G., A.V.G.-G., E.S., and F.L.; Writing—original draft, V.G. and E.P.-R.; Writing—review and editing, V.G., A.V.G.-G., E.S., M.Á.J.-C., and U.H. All authors have read and agreed to the published version of the manuscript. An earlier version of this manuscript constituted a chapter of the PhD thesis of the first author V.G.

**Funding:** This study has been supported by the project PAC08-0296-7771 (JCCM), and from INIA-MARM funds (INIA CC08-020). Elena Sotelo was a fellow from INIA. Elisa Perez Ramirez was a fellow of the National Research Council (CSIC) and Ana Valeria Gutierrez was a fellow of the JCCM.

**Institutional Review Board Statement:** The study was conducted according to the guidelines of the Declaration of Helsinki and approved by the Ethics and Animal Welfare Committee of the National Institute for Agricultural investigation and technology, Instituto Nacional de Investigación y Tecnología Agraria INIA (permit no. CEEA2008/013).

**Informed Consent Statement:** Not applicable.

**Data Availability Statement:** The data presented in this study are available in the different tables of this article.

**Acknowledgments:** We thank the personnel of the experimental farm of the University of Castilla-La Mancha "La Galiana" for their effort in this study. We would like to acknowledge the Junta de Comunidades de Castilla-La Mancha (JCCM) for their support. We thank Francisca Talavera Benitez for help with the preparation of tissue slides for immunohistochemistry and histopathology.

**Conflicts of Interest:** The authors declare no conflict of interest.

## References

1. Kramer, L.D.; Styer, L.M.; Ebel, G.D. A global perspective on the epidemiology of West Nile virus. *Annu. Rev. Entomol.* **2008**, *53*, 61–81. [CrossRef]
2. Bakonyi, T.; Haussig, J.M. West Nile virus keeps on moving up in Europe. *Euro. Surveill.* **2020**, *25*. [CrossRef] [PubMed]
3. European Centre for Disease Prevention and Control (ECDC). Available online: http://www.ecdc.europa.eu/en/healthtopics/west_nile_fever/West-Nile-fever-maps/pages/index.aspx (accessed on 10 December 2020).
4. Pauli, G.; Bauerfeind, U.; Blumel, J.; Burger, R.; Drosten, C.; Gröner, A.; Gürtler, L.; Heiden, M.; Hildebrandt, M.; Jansen, B.; et al. West Nile virus. *Transfus Med. Hemother.* **2013**, *40*, 265–284. [CrossRef]
5. Sotelo, E.; Fernández-Pinero, J.; Llorente, F.; Vázquez, A.; Moreno, A.; Agüero, M.; Cordioli, P.; Tenorio, A.; Jiménez-Clavero, M.A. Phylogenetic relationships of Western Mediterranean West Nile virus strains (1996–2010) using full-length genome sequences: Single or multiple introductions? *J. Gen Virol.* **2011**, *92*, 2512–2522. [CrossRef] [PubMed]
6. Malkinson, M.; Banet, C.; Weisman, Y.; Pokamunski, S.; King, R.; Drouet, M.T.; Deubel, V. Introduction of West Nile virus in the Middle East by migrating white storks. *Emerg. Infect Dis.* **2000**, *8*, 392–397. [CrossRef]
7. Höfle, U.; Blanco, J.M.; Crespo, E.; Naranjo, V.; Jiménez-Clavero, M.A.; Sánchez, A.; de la Fuente, J.; Gortazar, C. West Nile virus in the endangered Spanish imperial eagle. *Vet. Microbiol.* **2008**, *129*, 171–178. [CrossRef] [PubMed]
8. Jiménez-Clavero, M.A.; Sotelo, E.; Fernández-Pinero, J.; Llorente, F.; Blanco, J.M.; Rodriguez-Ramos, J.; Perez-Ramirez, E.; Höfle, U. West Nile virus in golden eagles, Spain, 2007. *Emerg. Infect. Dis.* **2008**, *14*, 1489–1491. [CrossRef]
9. Jourdain, E.; Schuffenecker, I.; Korimbocus, J.; Reynard, S.; Murri, S.; Kayser, Y.; Gauthier-Clerc, M.; Sabatier, P.; Zeller, H.G. West Nile virus in wild resident birds, Southern France, 2004. *Vector Borne Zoonotic Dis.* **2007**, *7*, 448–452. [CrossRef] [PubMed]
10. Monaco, F.; Lelli, R.; Teodori, L.; Pinoni, C.; Di Gennaro, A.; Polci, A.; Calistri, P.; Savini, G. Re-emergence of West Nile virus in Italy. *Zoonoses Public Health* **2010**, *57*, 476–486. [CrossRef]

11. Busquets, N.; Laranjo-González, M.; Soler, M.; Nicolás, O.; Rivas, R.; Talavera, S.; Villalba, R.; San Miguel, E.; Torner, N.; Aranda, C.; et al. Detection of West Nile virus lineage 2 in North-Eastern Spain (Catalonia). *Transbound. Emerg. Dis.* **2019**, *66*, 617–621. [CrossRef]
12. Valiakos, G.; Plavos, K.; Vontas, A.; Sofia, M.; Giannakopoulos, A.; Giannoulis, T.; Spyrou, V.; Tsokana, C.N.; Chatzopoulos, D.; Kantere, M.; et al. Phylogenetic Analysis of Bird-Virulent West Nile Virus Strain, Greece. *Emerg. Infect. Dis.* **2019**, *25*, 2323–2325. [CrossRef] [PubMed]
13. Hubálek, Z.; Kosina, M.; Rudolf, I.; Mendel, J.; Straková, P.; Tomešek, M. Mortality of Goshawks (Accipiter gentilis) Due to West Nile Virus Lineage 2. *Vector Borne Zoonotic Dis.* **2018**, *18*, 624–627. [CrossRef] [PubMed]
14. Del Amo, J.; Llorente, F.; Perez-Ramirez, E.; Soriguer, R.C.; Figuerola, J.; Nowotny, N.; Jiménez-Clavero, M.A. Experimental infection of house sparrows (*Passer domesticus*) with West Nile virus strains of lineages 1 and 2. *Vet. Microbiol.* **2014**, *172*, 542–547. [CrossRef] [PubMed]
15. Dridi, M.; Vangeluwe, D.; Lecollinet, S.; van den Berg, T.; Lambrecht, B. Experimental infection of Carrion crows (Corvus corone) with two European West Nile virus (WNV) strains. *Vet. Microbiol.* **2013**, *165*, 160–166. [CrossRef]
16. Lim, S.M.; Brault, A.C.; van Amerongen, G.; Sewbalaksing, V.D.; Osterhaus, A.D.; Martina, B.E.E.; Koraka, P. Susceptibility of European jackdaws (Corvus monedula) to experimental infection with lineage 1 and 2 West Nile viruses. *J. Gen. Virol.* **2014**, *95*, 1320–1329. [CrossRef]
17. Sotelo, E.; Gutiérrez-Guzmán, A.V.; del Amo, J.; Llorente, F.; El-Harrak, M.; Pérez-Ramírez, E.; Blanco, J.M.; Höfle, U.; Jiménez-Clavero, M.A. Pathogenicity of two recent Western Mediterranean West Nile virus isolates in a wild bird species indigenous to Southern Europe: The red-legged partridge. *Vet. Res.* **2011**, *42*, 11. [CrossRef]
18. Pérez-Ramírez, E.; Llorente, F.; Jiménez-Clavero, M.Á. Experimental infections of wild birds with West Nile virus. *Viruses* **2014**, *6*, 752–781. [CrossRef]
19. Blanco-Aguiar, J.A.; Virgós, E.; Villafuerte, R. Perdiz roja (*Alectoris rufa*). In *Atlas de las Aves Reproductoras de España*; Martí, R., del Moral, J.C., Eds.; DGCoNa-SEO: Madrid, Spain, 2003; pp. 212–213.
20. Schuffenecker, I.; Peyrefitte, C.N.; el Harrak, M.; Murri, S.; Leblond, A.; Zeller, H.G. West Nile virus in Morocco, 2003. *Emerg. Infect. Dis.* **2005**, *11*, 306–309. [CrossRef]
21. Sotelo, E.; Fernández-Pinero, J.; Llorente, F.; Agüero, M.; Hoefle, U.; Blanco, J.M.; Jiménze-Clavero, M.A. Characterization of West Nile virus isolates from Spain: New insights into the distinct West Nile virus eco-epidemiology in the Western Mediterranean. *Virology* **2009**, *395*, 289–297. [CrossRef]
22. Lim, S.M.; Koraka, P.; van Boheemen, S.; Roose, J.M.; Jaarsma, D.; van de Vijver, D.A.; Osterhaus, A.D.; Martina, B.E. Characterization of the mouse neuroinvasiveness of selected European strains of West Nile virus. *PLoS ONE* **2013**, *8*, e74575. [CrossRef]
23. Reisen, W.K.; Fang, Y.; Lothrop, H.D.; Martinez, V.M.; Wilson, J.; Oconnor, P.; Carney, R.; Cahoon-Young, B.; Shafii, M.; Brault, A.C. Overwintering of West Nile virus in Southern California. *J. Med. Entomol.* **2006**, *43*, 344–355. [CrossRef]
24. Wheeler, S.S.; Langevin, S.A.; Brault, A.C.; Woods, L.; Carroll, B.D.; Reisen, W.K. Detection of persistent West Nile virus RNA in experimentally and naturally infected avian hosts. *Am. J. Trop. Med. Hyg.* **2012**, *87*, 559–564. [CrossRef]
25. Steele, K.E.; Linn, M.J.; Schoepp, R.J.; Komar, N.; Geisbert, T.W.; Manduca, R.M.; Calle, P.P.; Raphael, B.L.; Clippinger, T.L.; Larsen, T.; et al. Pathology of fatal West Nile virus infections in native and exotic birds during the 1999 outbreak in New York City, New York. *Vet. Pathol.* **2000**, *37*, 208–224. [CrossRef] [PubMed]
26. Wünschmann, A.; Ziegler, A. West Nile virus-associated mortality events in domestic chukar partridges (*Alectoris chukar*) and domestic Impeyan pheasants (*Lophophorus impeyanus*). *Avian Dis.* **2006**, *50*, 456–459. [CrossRef] [PubMed]
27. Zhang, Z.; Wilson, F.; Read, R.; Pace, L.; Zhang, S. Detection and characterization of naturally acquired West Nile virus infection in a female wild turkey. *J. Vet. Diagn Invest.* **2006**, *18*, 204–208. [CrossRef] [PubMed]
28. Eckstrand, C.D.; Woods, L.W.; Diab, S.S.; Crossley, B.M.; Giannitti, F. Diagnostic Exercise: High Mortality in a Flock of Chukar Partridge Chicks (*Alectoris chukar*) in California. *Vet. Pathol.* **2015**, *52*, 189–192. [CrossRef]
29. Nemeth, N.M.; Bosco-Lauth, A.M.; Williams, L.M.; Bowen, R.A.; Brown, J.D. West Nile Virus Infection in Ruffed Grouse (*Bonasa umbellus*): Experimental Infection and Protective Effects of Vaccination. *Vet. Pathol.* **2017**, *54*, 901–911. [CrossRef]
30. Byas, A.D.; Ebel, G.D. Comparative Pathology of West Nile Virus in Humans and Non-Human Animals. *Pathogens* **2020**, *9*, 48. [CrossRef]
31. Gancz, A.Y.; Smith, D.A.; Barker, I.K.; Lindsay, R.; Hunter, B. Pathology and tissue distribution of West Nile virus in North American owls (family: Strigidae). *Avian Pathol.* **2006**, *35*, 17–29. [CrossRef]
32. Shirafuji, H.; Kanehira, K.; Kubo, M.; Shibahara, T.; Kamio, T. Experimental West Nile virus infection in jungle crows (*Corvus macrorhynchos*). *Am. J. Trop. Med. Hyg.* **2008**, *78*, 838–842. [CrossRef]
33. Wünschmann, A.; Shivers, J.; Bender, J.; Carroll, L.; Fuller, S.; Saggese, M.; van Wettere, A.; Redig, P. Pathologic and immunohistochemical findings in goshawks (*Accipiter gentilis*) and great horned owls (*Bubo virginianus*) naturally infected with West Nile virus. *Avian Dis.* **2005**, *49*, 252–259. [CrossRef] [PubMed]
34. Banet-Noach, C.; Simanov, L.; Malkinson, M. Direct (non-vector) transmission of West Nile virus in geese. *Avian Pathol.* **2003**, *32*, 489–494. [CrossRef] [PubMed]
35. Nemeth, N.M.; Kratz, G.E.; Bates, R.; Scherpelz, J.A.; Bowen, R.A.; Komar, N. Clinical evaluation and outcomes of naturally acquired West Nile virus infection in raptors. *J. Zoo Wildl. Med.* **2009**, *40*, 51–63. [PubMed]

36. Saito, E.K.; Sileo, L.; Green, D.E.; Meteyer, C.U.; McLaughlin, G.S.; Converse, K.A.; Docherty, D.E. Raptor mortality due to West Nile virus in the United States, 2002. *J. Wildl. Dis.* **2009**, *43*, 206–213. [CrossRef]
37. Busquets, N.; Bertran, K.; Costa, T.P.; Rivas, R.; de la Fuente, J.G.; Villalba, R.; Solanes, D.; Bensaid, A.; Majó, N.; Pagès, N. Experimental West Nile virus infection in Gyr-Saker hybrid falcons. *Vector Borne Zoonotic Dis.* **2012**, *12*, 482–489. [CrossRef]
38. Lim, S.M.; Koraka, P.; Osterhaus, A.D.; Martina, B.E. West Nile virus: Immunity and pathogenesis. *Viruses* **2011**, *3*, 811–828. [CrossRef]
39. Shrestha, B.; Gottlieb, D.; Diamond, M.S. Infection and injury of neurons by West Nile encephalitis virus. *J. Virol.* **2003**, *77*, 13203–13213. [CrossRef]
40. Feyer, S.; Bartenschlager, F.; Bertram, C.A.; Ziegler, U.; Fast, C.; Klopfleisch, R.; Müller, K. Clinical, pathological and virological aspects of fatal West Nile virus infections in ten free-ranging goshawks (*Accipiter gentilis*) in Germany. *Transbound. Emerg. Dis.* **2021**, *68*, 907–919. [CrossRef]
41. Pérez-Ramírez, E.; Llorente, F.; Del Amo, J.; Nowotny, N.; Jiménez-Clavero, M.Á. Susceptibility and role as competent host of the red-legged partridge after infection with lineage 1 and 2 West Nile virus isolates of Mediterranean and Central European origin. *Vet. Microbiol.* **2018**, *222*, 39–45. [CrossRef]
42. Montgomery, D.L. Astrocytes: Form, functions, and roles in disease. *Vet. Pathol.* **1994**, *31*, 145–167. [CrossRef]
43. Bréhin, A.C.; Mouries, J.; Frenkiel, M.P.; Dadaglio, G.; Despres, P.; Lafon, M.; Couderc, T. Dynamics of immune cell recruitment during West Nile encephalitis and identification of a new CD19+B220-BST-2+ leukocyte population. *J. Immunol.* **2008**, *180*, 6760–6767. [CrossRef] [PubMed]
44. Hussmann, K.L.; Samuel, M.A.; Kim, K.S.; Diamond, M.S.; Fredericksen, B.L. Differential replication of pathogenic and non-pathogenic strains of West Nile virus within astrocytes. *J. Virol.* **2013**, *87*, 2814–2822. [CrossRef]
45. Kelley, T.W.; Prayson, R.A.; Ruiz, A.I.; Isada, C.M.; Gordon, S.M. The neuropathology of West Nile virus meningoencephalitis. A report of two cases and review of the literature. *Am. J. Clin. Pathol.* **2003**, *119*, 749–753. [CrossRef] [PubMed]
46. Maximova, O.A.; Faucette, L.J.; Ward, J.M.; Murphy, B.R.; Pletnev, A.G. Cellular inflammatory response to flaviviruses in the central nervous system of a primate host. *J. Histochem. Cytochem.* **2009**, *57*, 973–989. [CrossRef]
47. Dorries, R. The role of T-cell-mediated mechanisms in virus infections of the nervous system. *Curr. Top Microbiol. Immunol.* **2001**, *253*, 219–245. [PubMed]
48. Sampson, B.A.; Armbrustmacher, V. West Nile encephalitis: The neuropathology of four fatalities. *Ann. N. Y. Acad. Sci.* **2001**, *951*, 172–178. [CrossRef]
49. Shrestha, B.; Diamond, M.S. Role of CD8+ T cells in control of West Nile virus infection. *J. Virol.* **2004**, *78*, 8312–8321. [CrossRef] [PubMed]
50. Sitati, E.M.; Diamond, M.S. CD4+ T-cell responses are required for clearance of West Nile virus from the central nervous system. *J. Virol.* **2006**, *80*, 12060–12069. [CrossRef]
51. Wang, Y.; Lobigs, M.; Lee, E.; Mullbacher, A. CD8+ T cells mediate recovery and immunopathology in West Nile virus encephalitis. *J. Virol.* **2003**, *77*, 13323–13334. [CrossRef] [PubMed]
52. Brault, A.C.; Huang, C.Y.H.; Langevin, S.A.; Kinney, R.M.; Bowen, R.A.; Ramey, W.N.; Panella, N.A.; Holmes, E.C.; Powers, A.M.; Miller, B.R. A single positively selected West Nile viral mutation confers increased virogenesis in American crows. *Nat. Genet.* **2007**, *39*, 1162–1166. [CrossRef]
53. Brault, A.C.; Langevin, S.A.; Ramey, W.N.; Fang, Y.; Beasley, D.W.; Barker, C.M.; Sanders, T.A.; Reisen, W.K.; Barrett, A.D.T.; Bowen, R.A. Reduced avian virulence and viremia of West Nile virus isolates from Mexico and Texas. *Am. J. Trop. Med. Hyg.* **2011**, *85*, 758–767. [CrossRef]
54. Davis, C.T.; Ebel, G.D.; Lanciotti, R.S.; Brault, A.C.; Guzman, H.; Siirin, M.; Lambert, A.; Parsons, R.E.; Beasley, D.W.C.; Novak, R.J.; et al. Phylogenetic analysis of North American West Nile virus isolates, 2001–2004, evidence for the emergence of a dominant genotype. *Virology* **2005**, *342*, 252–265. [CrossRef] [PubMed]
55. Fiacre, L.; Pagès, N.; Albina, E.; Richardson, J.; Lecollinet, S.; Gonzalez, G. Molecular Determinants of West Nile Virus Virulence and Pathogenesis in Vertebrate and Invertebrate Hosts. *Int. J. Mol. Sci.* **2020**, *21*, 9117. [CrossRef] [PubMed]
56. Chambers, T.J.; Diamond, M.S. Pathogenesis of flavivirus encephalitis. *Adv. Virus Res.* **2003**, *60*, 273–342. [PubMed]
57. Beasley, D.W.; Whiteman, M.C.; Zhang, S.; Huang, C.Y.; Schneider, B.S.; Smith, D.R.; Gromowski, G.D.; Higgs, S.; Kinney, R.M.; Barrett, A.D.T. Envelope protein glycosylation status influences mouse neuroinvasion phenotype of genetic lineage 1 West Nile virus strains. *J. Virol.* **2005**, *79*, 8339–8347. [CrossRef]
58. Brault, A.C.; Langevin, S.A.; Bowen, R.A.; Panella, N.A.; Biggerstaff, B.J.; Miller, B.R.; Komar, N. Differential virulence of West Nile strains for American crows. *Emerg. Infect. Dis.* **2004**, *10*, 2161–2168. [CrossRef]
59. Gamino, V.; Gutiérrez-Guzmán, A.V.; Fernández-de-Mera, I.G.; Ortíz, J.A.; Durán-Martín, M.; de la Fuente, J.; Gortázar, C.; Hófle, C. Natural Bagaza virus infection in game birds in southern Spain. *Vet. Res.* **2012**, *43*, 65. [CrossRef]

*pathogens*

MDPI

*Article*

# Evaluation of West Nile Virus Diagnostic Capacities in Veterinary Laboratories of the Mediterranean and Black Sea Regions

**Elisa Pérez-Ramírez [1,*,†], Cristina Cano-Gómez [1,†], Francisco Llorente [1], Ani Vodica [2], Ljubiša Veljović [3], Natela Toklikishvilli [4], Kurtesh Sherifi [5], Soufien Sghaier [6], Amel Omani [7], Aida Kustura [8], Kiril Krstevski [9], Ilke Karayel-Hacioglu [10], Naglaa Mohamed Hagag [11], Jeanne El Hage [12], Hasmik Davdyan [13], Mohd Saddam Bintarif [14], Bojan Adzic [15], Nabil Abouchoaib [16], Miguel Ángel Jiménez-Clavero [1,17] and Jovita Fernández-Pinero [1]**

1 Centro de Investigación en Sanidad Animal, Instituto Nacional de Investigación y Tecnología Agraria y Alimentaria (INIA-CISA), 28130 Valdeolmos, Spain
2 Department of Animal Health, Food Safety and Veterinary Institute, Tirana, Albania
3 Virology Department, Scientific Institute of Veterinary Medicine of Serbia, 11000 Belgrade, Serbia
4 Laboratory of Virology and Molecular Biology, LEPL State Laboratory of Agriculture (SLA), 0159 Tbilisi, Georgia
5 Department of Veterinary Medicine, Faculty of Agriculture and Veterinary Sciences, University of Prishtina "Hasan Pristhina", 10000 Prishtine, Kosovo
6 Virology Department, Institute of Veterinary Research of Tunisia, 1006 Tunis, Tunisia
7 Laboratoire Central Vétérinaire d'Alger, Institut National de la Médecine Vétérinaire, Algiers, Algeria
8 Veterinary Faculty, University of Sarajevo, 71000 Sarajevo, Bosnia and Herzegovina
9 Faculty of Veterinary Medicine, Ss. Cyril and Methodius University, 1000 Skopje, North Macedonia
10 Virology Department, Faculty of Veterinary Medicine, Ankara University, 06110 Ankara, Turkey
11 Animal Health Research Institute, Dokki 12618, Egypt
12 Animal Health Laboratory, Lebanese Agricultural Research Institute, 90-1064 Fanar, Lebanon
13 Republican Veterinary-Sanitary and Phytosanitary Center of Laboratory Services SNCO, Yerevan, Armenia
14 Animal Wealth Laboratory Sector, Ministry of Agriculture, Amman, Jordan
15 Diagnostic Veterinary Laboratory, 81000 Podgorica, Montenegro
16 Casablanca Regional Research and Analysis Laboratory of National Office of Sanitary Safety and Food Products (ONSSA), Nouaceur, 20 000 Casablanca, Morocco
17 CIBER Epidemiología y Salud Pública (CIBERESP), 28029 Madrid, Spain
* Correspondence: elisaperezramirez@gmail.com
† Equally contributed.

Received: 4 November 2020; Accepted: 8 December 2020; Published: 11 December 2020

**Abstract:** The increasing incidence of West Nile virus (WNV) in the Euro-Mediterranean area warrants the implementation of effective surveillance programs in animals. A crucial step in the fight against the disease is the evaluation of the capacity of the veterinary labs to accurately detect the infection in animal populations. In this context, the animal virology network of the MediLabSecure project organized an external quality assessment (EQA) to evaluate the WNV molecular and serological diagnostic capacities of beneficiary veterinary labs. Laboratories from 17 Mediterranean and Black Sea countries participated. The results of the triplex real time RT-PCR for simultaneous detection and differentiation of WNV lineage 1 (L1), lineage 2 (L2) and Usutu virus (USUV) were highly satisfactory, especially for L1 and L2, with detection rates of 97.9% and 100%, respectively. For USUV, 75% of the labs reported correct results. More limitations were observed for the generic detection of flaviviruses using conventional reverse-transcription polymerase chain reaction (RT-PCR), since only 46.1% reported correct results in the whole panel. As regards the serological panel, the results were excellent for the generic detection of WNV antibodies. More variability was observed for the specific detection of IgM antibodies with a higher percentage of incorrect results mainly in samples with low titers. This EQA provides a good overview of the WNV (and USUV) diagnostic performance of the involved veterinary labs and demonstrates that the implemented training program was successful in upgrading their diagnostic capacities.

**Keywords:** West Nile virus (WNV); Usutu virus (USUV); flavivirus; external quality assessment (EQA); MediLabSecure; diagnostics; PCR; ELISA

---

## 1. Introduction

West Nile virus (WNV) is an enveloped spherical, single-stranded positive-sense RNA virus belonging to the *Flaviviridae* family [1]. It is maintained in nature in an enzootic cycle involving ornithophilic mosquitoes (mainly *Culex*) as transmission vectors and certain birds as reservoir hosts. Spill-over from this cycle occasionally results in severe outbreaks in horses and humans that are considered dead-end hosts, which means that they cannot transmit the virus to feeding mosquitoes due to their low and transient viremia [2].

Although most human infections are asymptomatic, in some instances (~20%) the virus can cause a febrile syndrome (WNV fever). Around 1% of the cases progress to severe neuro-invasive disease with a fatality rate of around 10% [3]. In horses, the infection is generally asymptomatic, but approximately 10% of the infected animals develop neurological symptoms such as ataxia, limb paralysis, skin fasciculation and muscle tremors [4]. In birds, the pathogenic potential greatly differs among species and also depends on the viral strain. Passerine birds (especially corvids) and some raptor species are particularly susceptible to WNV infection [5].

Phylogenetic studies have revealed the existence of at least seven genetic lineages, of which lineages 1 (L1) and 2 (L2) are the most widespread and relevant for human and animal health [6]. Circulation of the virus has been regularly reported in wide areas of the Euro-Mediterranean area since 1998 [7]. However, in recent years, the virus has dramatically expanded, with an upsurge in the number and incidence of outbreaks in humans and animals caused by both L1 and L2 strains. In fact, in 2018 the transmission season started earlier and the number of human autochthonous cases reported in EU and neighboring countries (2083 cases) exceeded the global number from the previous seven years. In horses, 285 cases were notified, which represents an increase of 30% in comparison with 2017 [8]. However, epidemiological data on WNV from Black Sea countries is scarce and therefore surveillance efforts in this region needs to be harmonized with those implemented in other countries.

The epidemiological situation in the Mediterranean basin is complex because, apart from WNV L1 and L2, other flaviviruses circulate in overlapping areas [6,9]. This is the case with Usutu virus (USUV), a zoonotic arbovirus closely related to WNV that has spread throughout Europe since 2001, when it was first detected in Austria [10]. Nowadays many countries have reported the presence of both viruses in mosquitos, birds and humans [9,11]. Of the countries participating in this EQA, in at least three (Serbia, Tunisia and Morocco) the co-circulation of both viruses has been demonstrated [12–14].

WNV diagnostic methods include virus isolation, reverse-transcription polymerase chain reaction (RT-PCR) and serological tests. Isolation procedures are laborious and require biosecurity level 3 (BSL-3) facilities. By contrast, molecular methods, such as RT-PCR, and particularly real-time RT-PCR (RRT-PCR), can be easily applied in basic laboratories, are fast and sensitive, enabling timely detection and early outbreak response [6]. In the current context, RRT-PCRs that allow for simultaneous detection of WNV L1, L2 and other flaviviruses such as USUV are extremely useful for outbreak investigations and epidemiological studies, maximizing the information obtained from each sample [15].

Among the antibody detection tools, ELISA tests are the most widely used with several commercial kits available. However, a relevant limitation of WNV ELISA tests is the cross-reactivity of antibodies raised against different flaviviruses that can lead to diagnostic misinterpretations [6,16]. To confirm WNV infection, the gold standard method is virus neutralization (VNT) that enables differential diagnosis by titration of neutralizing antibodies in parallel against different flaviviruses that could cross-react in serological tests. Nevertheless, VNT is time-consuming and has to be performed in BSL-3 labs.

Prevention and control efforts substantially rely on effective surveillance of the infection in animals and vectors that can act as early warning triggers [17]. The implementation of locally adapted surveillance systems in birds, horses and mosquitos and the upgrade of the diagnostic capacities of veterinary laboratories is crucial to fight the disease.

MediLabSecure is an EU-funded project whose main objective is to create a framework for collaboration to promote arbovirus surveillance under a One Health approach in 19 countries of the Mediterranean and Black Sea regions [18,19]. Since the beginning of the project in 2014, the MediLabSecure animal virology network has implemented numerous actions to enhance capacity building of veterinary labs to face health threats caused by emerging arboviruses.

In a specific questionnaire delivered to identify the priorities of the beneficiary countries in the field of arboviral diseases, all the labs recognized WNV as a common health priority in the region. A training curriculum was implemented to improve diagnostic performance of the veterinary labs for this pathogen, including two diagnostic workshops (molecular and serological diagnosis) that were organized in 2015 and 2016 at Centro de Investigación en Sanidad Animal (INIA-CISA) (Madrid, Spain). After these training sessions, an external quality assessment (EQA) was organized between October 2016 and March 2017 to evaluate the degree of learning and the capacity of the labs to incorporate the molecular and serological techniques into their routine diagnostic activities. In this study we report the results of this inter-laboratory trial and provide relevant information about the current WNV (and USUV) diagnostic capacities of veterinary labs in the Mediterranean and Black Sea regions. This exercise also enabled an extensive reproducibility assessment of the recommended tests for WNV and USUV diagnostics.

## 2. Results

Seventeen laboratories submitted results, representing 17 countries from the Mediterranean and Black Sea regions (Figure 1).

### *2.1. Virus Genome Detection*

A total of 13 datasets were received from 13 labs (76.4% of response) for the generic detection of flaviviruses. The results of all labs are shown in Table 1. Four labs did not carry out this technique due to lack of specific equipment to perform conventional RT-PCR assays and were therefore unable to detect the flavivirus positive sample of the panel corresponding to Japanese encephalitis virus (JEV).

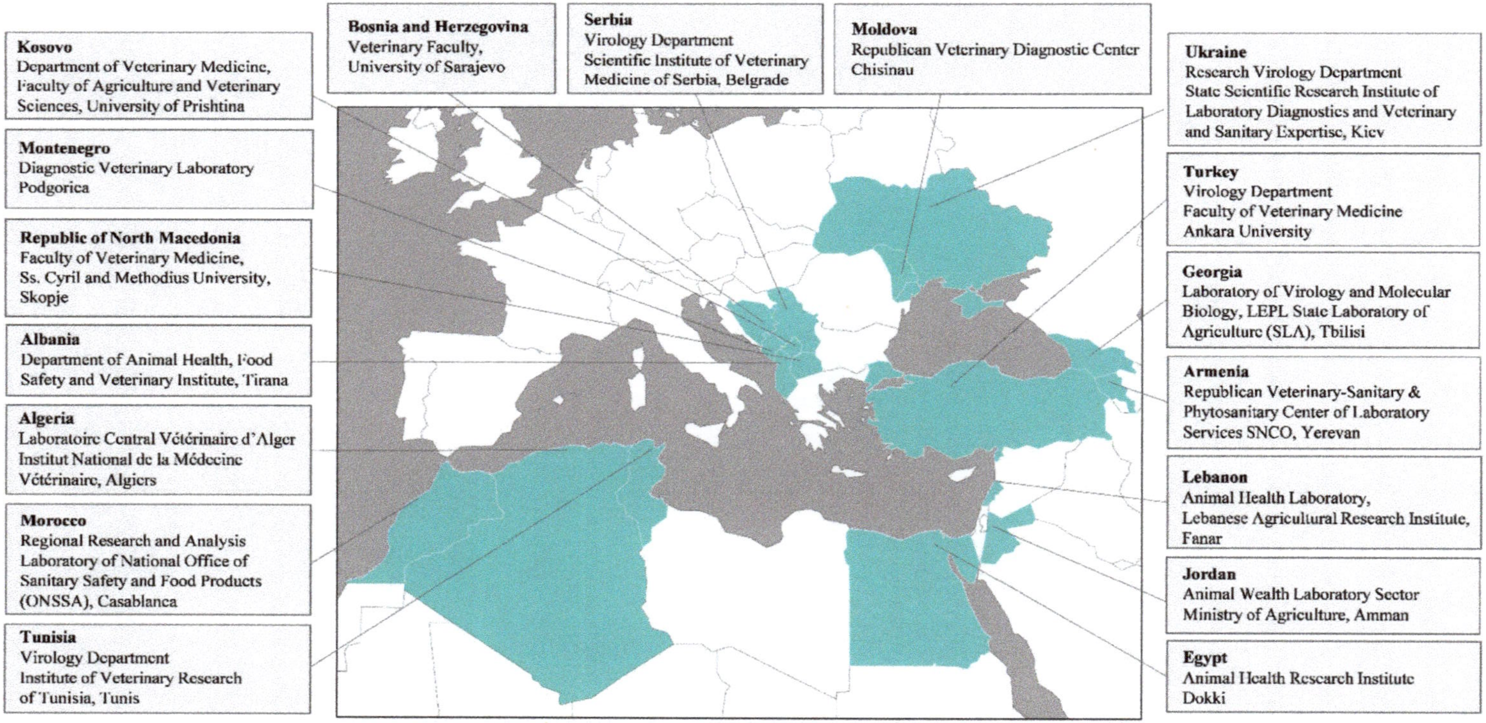


**Figure 1.** Participant laboratories. Obtained from [20].

**Table 1.** Results of the conventional reverse-transcription polymerase chain reaction (RT-PCR) for generic detection of flaviviruses.

| **Virus** | **West Nile Virus (WNV) L1** | | **WNV L2** | | **Usutu Virus (USUV)** | **Japanese Encephalitis Virus (JEV)** | **WNV L1/ USUV** | **WNV L2/ USUV** | – | – | |
|---|---|---|---|---|---|---|---|---|---|---|---|
| Strain | SP07 | | AUS08 | | USU11 | Nakayama | SP07/ USU11 | AUS08/USU11 | – | – | |
| Matrix (species) | Kidney (pheasant) | Liver (pheasant) | Serum (horse) | | Blood (horse) | Serum (horse) | Heart (partridge) | Serum (horse) | Brain (horse) | Heart (pheasant) | |
| Dilution | $10^{-2}$ | $10^{-3}$ | $10^{-1}$ | $10^{-3}$ | $10^{-3}$ | $10^{-2}$ | $10^{-4}/10^{-4}$ | $10^{-2}/10^{-4}$ | – | – | |
| Sample ID | W10 | W1 | W2 | W5 | W6 | W9 | W3 | W8 | W4 | W7 | |
| Reference value | Weak + | Weak + | Strong + | + | + | + | Weak + | Strong + | - | - | |
| Laboratory | | | | | | | | | | | % of correct results (by lab) |
| 1 | NA | NA | NA | NA | NA | NA | NA | NA | NA | NA | – |
| 2 | - | - | + | + | + | + | - | + | - | - | 70 |
| 3 | + | - | + | + | + | + | - | + | - | - | 80 |
| 4 | NA | NA | NA | NA | NA | NA | NA | NA | NA | NA | – |
| 5 | + | + | + | + | + | + | + | + | - | - | 100 |
| 6 | NA | NA | NA | NA | NA | NA | NA | NA | NA | NA | – |
| 7 | + | + | + | + | + | + | + | + | - | - | 100 |
| 8 | + | + | + | + | + | - | - | + | + | + | 60 |
| 9 | + | - | + | - | - | + | - | + | - | - | 60 |
| 10 | - | - | + | + | + | + | - | + | - | - | 70 |
| 11 | + | + | + | + | + | + | + | + | - | - | 100 |
| 12 | + | + | + | + | + | + | + | + | - | - | 100 |
| 13 | - | + | + | - | + | - | - | + | - | + | 50 |
| 14 | + | + | + | + | + | + | + | + | - | - | 100 |
| 15 | + | + | + | + | + | + | + | + | - | - | 100 |
| 16 | - | - | + | + | + | - | - | + | - | - | 60 |
| 17 | NA | NA | NA | NA | NA | NA | NA | NA | NA | NA | – |
| % of correct results (by sample) | 69.2 | 61.5 | 100 | 84.6 | 92.3 | 76.9 | 46.2 | 100 | 92.3 | 84.6 | |

Blue: false negative results; Red: false positive results; NA: not analyzed.

Out of the 13 datasets, only 6 (46.1%) were 100% correct, i.e., identified as positive the eight samples containing WNV L1, L2, USUV or JEV and as negative the two negative samples of the panel. However, it should be noted that for three of the positive samples, weak positive results (weak bands) were expected. The two samples (W2 and W8) with expected strong positive bands were correctly identified by all the labs (Table 1). Three false positive results were reported by two labs.

For the specific detection of WNV and USUV by RRT-PCR, we received 18 datasets from the 17 participating labs (100% of response), including one double dataset from lab #16 that used an alternative method [21] apart from the recommended one. The results of the labs using the recommended method are shown in Table 2. Overall, the results of the triplex RRT-PCR were very good, with the exception of one laboratory (#1) that reported incorrect results in the whole panel (Table 2). If we exclude the results of this lab from the global analysis of the EQA, out of 16 labs, 12 (75%) reported 100% concordant results, which means that they correctly identified WNV L1, L2 and USUV in all the positive samples (including co-infections).

However, a technical limitation was observed in the application of the triplex RRT-PCR in two labs (#8 and #12), because the Cy5 fluorescent channel necessary for the detection of USUV was lacking in the available real-time thermocyclers. Interestingly, one of these labs was able to overcome this limitation by applying an alternative conventional RT-PCR for the specific detection of USUV [22], obtaining 100% correct identification of USUV positive samples.

In general, highly satisfactory results were obtained for WNV L1 and L2 while more difficulties were observed for the correct identification of USUV positive samples (W3, W6 and W8) as shown in Table 2. In fact, out of the eight false negative results reported by four labs, seven corresponded to USUV and one to WNV L1.

Excluding results from lab #1, co-infections were successfully detected in 87.5% of cases for sample W3 (WNV L1+USUV) and in 81.2% of cases for sample W8 (WNV L2+USUV). The sample containing a related flavivirus (JEV) was correctly identified as negative by all the labs. Overall, the reported Ct values for the three viruses were in line with the reference values, except for labs #3 and #10 that reported lower Ct values for WNV L2 and labs #2 and #17 that reported higher Ct values than expected for WNV L2 and USUV in three samples. With regard to the negative samples, and excluding lab #1, only one laboratory reported a false positive result. Differences in qualitative results or Ct values were not attributable to a specific thermocycler.

Lab #16 correctly identified the WNV positive samples of the panel using a pair of alternative RT-PCR methods for specific detection of WNV L1 and L2 [21].

**Table 2.** Results of the triplex RRT-PCR for detection and differentiation of WNV L1, L2 and USUV.

| **Virus** | | **WNV L1** | | **WNV L2** | | **USUV** | **JEV** | **WNV L1/USUV** | **WNV L2/USUV** | – | – | | |
|---|---|---|---|---|---|---|---|---|---|---|---|---|---|
| Strain | | SP07 | | AUS08 | | USU11 | Nakayama | SP07/ USU11 | AUS08/ USU11 | – | – | | |
| Matrix (species) | | Kidney (pheasant) | Liver (pheasant) | Serum (horse) | | Blood (horse) | Serum (horse) | Heart (partridge) | Serum (horse) | Brain (horse) | Heart (pheasant) | | |
| Dilution | | $10^{-2}$ | $10^{-3}$ | $10^{-1}$ | $10^{-3}$ | $10^{-3}$ | $10^{-2}$ | $10^{-4}/10^{-4}$ | $10^{-2}/10^{-4}$ | – | – | | |
| Sample ID | | W10 | W1 | W2 | W5 | W6 | W9 | W3 | W8 | W4 | W7 | | |
| Reference Ct value | WNV L1 (FAM) | 30.68 ± 0.88 | 33.60 ± 0.98 | No Ct | No Ct | No Ct | No Ct | 32.01 ± 0.41 | No Ct | No Ct | No Ct | | |
| | WNV L2 (VIC) | No Ct | No Ct | 27.13 ± 0.57 | 34.28 ± 0.13 | No Ct | No Ct | No Ct | 31.34 ± 0.25 | No Ct | No Ct | | |
| | USUV (Cy5) | No Ct | No Ct | No Ct | No Ct | 29.24 ± 0.46 | No Ct | 32.1 ± 0.58 | 33.23 ± 0.31 | No Ct | No Ct | | |
| Lab | | | | | | | | | | | | % of correct results for each virus (by lab) | % of correct overall results (by lab) † |
| 1 | WNV L1 (FAM) | No Ct | No Ct | 32.79 | No Ct | No Ct | No Ct | No Ct | No Ct | 31.98 | No Ct | 60 | 10 |
| | WNV L2 (VIC) | No Ct | No Ct | No Ct | No Ct | 34.59 | 30 | 24.8 | No Ct | No Ct | No Ct | 40 | |
| | USUV (Cy5) | No Ct | No Ct | No Ct | No Ct | No Ct | 38 | No Ct | No Ct | 33.53 | 36.2 | 50 | |
| 2 | WNV L1 (FAM) | 27.75 | 33.32 | No Ct | No Ct | No Ct | No Ct | 32.4 | No Ct | No Ct | No Ct | 100 | 100 |
| | WNV L2 (VIC) | No Ct | No Ct | 37.04 | 32.87 | No Ct | No Ct | No Ct | 30.71 | No Ct | No Ct | 100 | |
| | USUV (Cy5) | No Ct | No Ct | No Ct | No Ct | 37.57 | No Ct | 34.1 | 36.19 | No Ct | No Ct | 100 | |
| 3 | WNV L1 (FAM) | 30.15 | 31.71 | No Ct | No Ct | No Ct | No Ct | 30.31 | No Ct | No Ct | No Ct | 100 | 100 |
| | WNV L2 (VIC) | No Ct | No Ct | 21.66 | 26.66 | No Ct | No Ct | No Ct | 23.12 | No Ct | No Ct | 100 | |
| | USUV (Cy5) | No Ct | No Ct | No Ct | No Ct | 28.89 | No Ct | 28.58 | 30.24 | No Ct | No Ct | 100 | |

**Table 2.** *Cont.*

| | | | | | | | | | | | | | |
|---|---|---|---|---|---|---|---|---|---|---|---|---|---|
| 4 | WNV L1 (FAM) | 28.28 | 31.07 | No Ct | No Ct | No Ct | No Ct | 30.44 | No Ct | No Ct | No Ct | 100 | 100 |
| | WNV L2 (VIC) | No Ct | No Ct | 24.53 | 32.55 | No Ct | No Ct | No Ct | 28.01 | No Ct | No Ct | 100 | |
| | USUV (Cy5) | No Ct | No Ct | No Ct | No Ct | 24.8 | No Ct | 27.45 | 28.02 | No Ct | No Ct | 100 | |
| 5 | WNV L1 (FAM) | 29.44 | 32.74 | No Ct | No Ct | No Ct | No Ct | 30.52 | No Ct | No Ct | No Ct | 100 | 100 |
| | WNV L2 (VIC) | No Ct | No Ct | 24.67 | 32.65 | No Ct | No Ct | No Ct | 28.06 | No Ct | No Ct | 100 | |
| | USUV (Cy5) | No Ct | No Ct | No Ct | No Ct | 31.89 | No Ct | 36.92 | 33.74 | No Ct | No Ct | 100 | |
| 6 | WNV L1 (FAM) | 31.6 | 36.8 | No Ct | No Ct | No Ct | No Ct | 34.5 | No Ct | No Ct | No Ct | 100 | 70 |
| | WNV L2 (VIC) | No Ct | No Ct | 30.2 | 38.7 | No Ct | No Ct | No Ct | 33.9 | No Ct | No Ct | 100 | |
| | USUV (Cy5) | No Ct | No Ct | No Ct | No Ct | No Ct | No Ct | No Ct | No Ct | No Ct | No Ct | 70 | |
| 7 | WNV L1 (FAM) | 31.84 | 31.88 | No Ct | No Ct | No Ct | No Ct | 32.3 | No Ct | No Ct | No Ct | 100 | 70 |
| | WNV L2 (VIC) | No Ct | No Ct | 25.54 | 32.78 | No Ct | No Ct | No Ct | 29.87 | No Ct | No Ct | 100 | |
| | USUV (Cy5) | No Ct | No Ct | No Ct | No Ct | No Ct | No Ct | No Ct | No Ct | No Ct | No Ct | 70 | |
| 8 | WNV L1 (FAM) | 31.2 | No Ct | No Ct | No Ct | No Ct | No Ct | 33.46 | No Ct | No Ct | No Ct | 90 | 90 |
| | WNV L2 (VIC) | No Ct | No Ct | 31 | 35.5 | No Ct | No Ct | No Ct | 32.5 | No Ct | No Ct | 100 | |
| | USUV (Cy5) | NA | NA | NA | NA | NA | NA | NA | NA | NA | NA | - | |
| 9 | WNV L1 (FAM) | 29 | 32 | No Ct | No Ct | No Ct | No Ct | 29 | No Ct | No Ct | No Ct | 100 | 100 |
| | WNV L2 (VIC) | No Ct | No Ct | 27 | 37 | No Ct | No Ct | No Ct | 32 | No Ct | No Ct | 100 | |
| | USUV (Cy5) | No Ct | No Ct | No Ct | No Ct | 28 | No Ct | 32 | 34 | No Ct | No Ct | 100 | |

**Table 2.** *Cont.*

| | | | | | | | | | | | | | |
|---|---|---|---|---|---|---|---|---|---|---|---|---|---|
| 10 | WNV L1 (FAM) | 28.19 | 30.76 | No Ct | No Ct | No Ct | No Ct | 29.59 | No Ct | No Ct | No Ct | 100 | 100 |
| | WNV L2 (VIC) | No Ct | No Ct | 22.11 | 29.8 | No Ct | No Ct | No Ct | 26.78 | No Ct | No Ct | 100 | |
| | USUV (Cy5) | No Ct | No Ct | No Ct | No Ct | 28.39 | No Ct | 30.38 | 30.97 | No Ct | No Ct | 100 | |
| 11 | WNV L1 (FAM) | 29.03 | 31.57 | No Ct | No Ct | No Ct | No Ct | 30.96 | No Ct | No Ct | No Ct | 100 | 100 |
| | WNV L2 (VIC) | No Ct | No Ct | 25.46 | 32.65 | No Ct | No Ct | No Ct | 28.75 | No Ct | No Ct | 100 | |
| | USUV (Cy5) | No Ct | No Ct | No Ct | No Ct | 28.81 | No Ct | 29.46 | 30.95 | No Ct | No Ct | 100 | |
| 12 | WNV L1 (FAM) | 32.32 | 34.5 | No Ct | No Ct | No Ct | No Ct | 32.09 | No Ct | No Ct | No Ct | 100 | 100 |
| | WNV L2 (VIC) | No Ct | No Ct | 29.39 | 35.69 | No Ct | No Ct | No Ct | 32.11 | No Ct | No Ct | 100 | |
| | USUV (Cy5) | NA | NA | NA | NA | NA | NA | NA | NA | NA | NA | - | |
| 13 | WNV L1 (FAM) | 32.38 | 34.03 | No Ct | No Ct | No Ct | No Ct | 33.69 | No Ct | No Ct | No Ct | 100 | 100 |
| | WNV L2 (VIC) | No Ct | No Ct | 27.8 | 36.3 | No Ct | No Ct | No Ct | 31.45 | No Ct | No Ct | 100 | |
| | USUV (Cy5) | No Ct | No Ct | No Ct | No Ct | 36 | No Ct | 36.7 | 37.7 | No Ct | No Ct | 100 | |
| 14 | WNV L1 (FAM) | 28 | 32 | No Ct | No Ct | No Ct | No Ct | 31.6 | No Ct | No Ct | No Ct | 100 | 100 |
| | WNV L2 (VIC) | No Ct | No Ct | 26.4 | 30.6 | No Ct | No Ct | No Ct | 26.4 | No Ct | No Ct | 100 | |
| | USUV (Cy5) | No Ct | No Ct | No Ct | No Ct | 30.8 | No Ct | 32.5 | 32.8 | No Ct | No Ct | 100 | |
| 15 | WNV L1 (FAM) | 31.66 | 32.86 | No Ct | No Ct | No Ct | No Ct | 34.11 | No Ct | No Ct | No Ct | 100 | 100 |
| | WNV L2 (VIC) | No Ct | No Ct | 23.19 | 32.84 | No Ct | No Ct | No Ct | 27.41 | No Ct | No Ct | 100 | |
| | USUV (Cy5) | No Ct | No Ct | No Ct | No Ct | 29.45 | No Ct | 35.73 | 29.6 | No Ct | No Ct | 100 | |

**Table 2.** *Cont.*

| | | | | | | | | | | | | | |
|---|---|---|---|---|---|---|---|---|---|---|---|---|---|
| 16 | WNV L1 (FAM) | 32 | 31.9 | No Ct | No Ct | No Ct | No Ct | 31.2 | No Ct | No Ct | No Ct | 100 | 100 |
| | WNV L2 (VIC) | No Ct | No Ct | 24 | 33.5 | No Ct | No Ct | No Ct | 32.8 | No Ct | No Ct | 100 | |
| | USUV (Cy5) | No Ct | No Ct | No Ct | No Ct | 27.8 | No Ct | 32.6 | 28.2 | No Ct | No Ct | 100 | |
| 17 | WNV L1 (FAM) | 34 | 30 | No Ct | No Ct | No Ct | No Ct | 32 | No Ct | No Ct | No Ct | 100 | 80 |
| | WNV L2 (VIC) | No Ct | No Ct | 28 | 34 | No Ct | No Ct | No Ct | 37 | 32 | No Ct | 90 | |
| | USUV (Cy5) | No Ct | No Ct | No Ct | No Ct | 31 | No Ct | 28 | No Ct | No Ct | No Ct | 90 | |
| % of correct results (by sample) ‡ | 94.1 | 88.2 | 94.1 | 94.1 | 82.3 | 94.1 | 82.3 | 76.4 | 94.1 | 94.1 | 94.1 | | |

Blue: false negative results; Red: false positive results; NA: not analyzed due to lack of appropriate fluorophore (Cy5) † Percentage of samples where all the viruses have been correctly identified; ‡ Percentage of 100% correct results by sample (all the viruses of each sample have been correctly identified).

## 2.2. Antibody Detection

A total of 17 datasets were received from the 17 labs (100% response) for the generic detection of WNV antibodies using commercial competition ELISAs. Sixteen labs used the recommended method (INgezim West Nile Compac ELISA) and one lab (#8) applied an alternative commercial kit (ID Screen West Nile Competition Multi-species, IDvet).

All the labs that used the Ingenasa kit reported 100% concordant results, except for one negative sample that was assigned as doubtful by lab #7. The only lab that used the IDVet kit also reported 100% correct results.

For the specific detection of WNV IgM antibodies, 19 datasets were received from 17 labs (100% response). Sixteen labs used the recommended method (INgezim WNV IgM ELISA), and two (#14 and #16) also analyzed the panel with an alternative MAC ELISA, the ID Screen West Nile IgM Capture kit from IDvet. Laboratory #8 used this kit instead of the recommended one.

The results of the IgM antibody detection using the recommended method are presented in Table 3. Correct results in the whole panel were reported by 56.2% of the labs. It is important to note that, of the 4 IgM positive sera, two (W2 and W8) had low IgM antibody titers, which might explain the difficulties of some labs in identifying these samples. In fact, these sera displayed OD values close to the reference limits producing a higher percentage of false negative results (Table 3). The serum with high IgM antibody titer (W4) was successfully identified by all labs except one (#4). The six IgM negative sera were correctly assigned by all the labs except for one sample (W6) that was reported as IgM doubtful by lab #15.

As explained earlier, three labs used an alternative ELISA kit to analyze the panel. In all of them this kit failed to detect three out of the four IgM positive sera of the panel. Only the sample with high IgM antibody titer (W4) could be correctly identified by the IDVet IgM capture ELISA in the three labs.

**Table 3.** Results of the external quality assessment (EQA) for WNV IgM antibody detection using the recommended method (INgezim West Nile IgM ELISA).

| Serum Sample | Horse (past infection) | | Horse (recent infection) | | | | Non-infected horse | | Negative commercial horse serum | | |
|---|---|---|---|---|---|---|---|---|---|---|---|
| | W3 | W6 | W2 | W8 | W10 | W4 | W1 | W9 | W5 | W7 | |
| Reference OD values | 0.11 ± 0.05 | 0.25 ± 0.21 | 0.34 ± 0.04 | 0.44 ± 0.09 | 0.48 ± 0.12 | 1.82 ± 0.21 | 0.07 ± 0.03 | 0.12 ± 0.11 | 0.10± 0.11 | 0.05 ± 0.03 | |
| Reference qualitative result | - | - | D/+ | + | + | + | - | - | - | - | |
| Laboratory | | | | | | | | | | | % of correct results (by lab) |
| 1 | - | - | + | - | + | + | - | - | - | - | 90 |
| 2 | - | - | D | + | + | + | - | - | - | - | 100 |
| 3 | - | - | + | + | + | + | - | - | - | - | 100 |
| 4 | - | - | + | + | + | - | - | - | - | - | 90 |
| 5 | - | - | + | + | + | + | - | - | - | - | 100 |
| 6 | - | - | - | - | D | + | - | - | - | - | 70 |
| 7 | - | - | + | + | + | + | - | - | - | - | 100 |
| 8 | NA | NA | NA | NA | NA | NA | NA | NA | NA | NA | – |
| 9 | - | - | + | + | + | + | - | - | - | - | 100 |
| 10 | - | - | D | - | + | + | - | - | - | - | 90 |
| 11 | - | - | + | + | + | + | - | - | - | - | 100 |
| 12 | - | - | + | + | + | + | - | - | - | - | 100 |
| 13 | - | - | + | + | + | + | - | - | - | - | 100 |
| 14 | - | - | D | D | + | + | - | - | - | - | 90 |
| 15 | - | D | D | + | + | + | - | - | - | - | 90 |
| 16 | - | - | D | + | + | + | - | - | - | - | 100 |
| 17 | - | - | - | D | + | + | - | - | - | - | 80 |
| % of correct results (by sample) | 100 | 93.7 | 87.5 | 68.7 | 93.7 | 93.7 | 100 | 90 | 100 | 100 | |

Grey: incorrect result; NA: not analyzed; D: doubtful.

## 3. Discussion

Despite the increasing incidence of WNV in Europe and the Mediterranean area, few external quality assessments have been organized to evaluate the diagnostic capacities of labs, especially in the veterinary sector. In the human health sector, five EQAs were organized by the ENIVD and EVD-LabNet networks between 2006 and 2017 to evaluate the WNV diagnostic performance of human virology labs in different countries, mostly in Europe but also in Middle East and America [23–27]. In Italy, two EQAs were organized in 2010 and 2011 by the National Institute of Health to assess the capacity of the national blood transfusion centers to detect WNV genome on blood donations [28]. In the veterinary sector, only two inter-laboratory assays were carried out in 2010 and 2013 to evaluate the capacity of the National Reference labs for equine diseases in Europe and Morocco to detect WNV antibodies in horse sera [29].

The EQA we present here has a number of differential features with respect to prior WNV inter-laboratory trials. In the first place, the geographical coverage of the involved laboratories. In previous EQAs, most of the participant labs were European while in this case labs from 17 non-EU countries including North Africa, Balkans, Black Sea and Middle East regions have been involved. Of these, only one lab in Bosnia and Herzegovina had participated in the previous serology EQA organized by ANSES in 2013 [29]. Secondly, this EQA was organized as a final evaluation after a three year training period where the participant labs attended several hands-on workshops aimed at improving their diagnostic performance. Thirdly, as the diagnostic assay used for molecular detection of WNV is a triplex RRT-PCR that allows the simultaneous detection of USUV, this EQA also provides relevant data about the capacities of the labs to identify this emerging flavivirus. Last, but not least, the labs were provided with all the materials required to perform the recommended diagnostic assays. The objective was to facilitate as much as possible the participation of all beneficiary labs. Moreover, and to promote sustainability, positive extraction and PCR controls were also provided to be used as quality controls during this EQA but also to serve as reference material for the future diagnostic activities of the labs. Such material is otherwise difficult to obtain and was greatly appreciated by the participants.

An added value of this EQA is also the combination of molecular and serological methods since this integrated approach is essential for WNV surveillance in animal populations. Although previous EQAs had analyzed the performance of vet labs to detect WNV antibodies, this is the first international proficiency test to evaluate the molecular diagnostic capacities of animal diagnostic labs.

The panel for WNV genome detection consisted of 10 samples containing various concentrations of four different flavivirus strains: two WNV European strains representing lineages 1 and 2, one European USUV strain and the reference JEV Nakayama strain. Unlike previous WNV EQAs where the viruses were diluted in human plasma [23,27,28] or virus culture medium [24], in this case the viruses were spiked in different organs' homogenates, serum or blood from horses and birds to mimic as much as possible the clinical samples that the vet labs would analyze during real surveillance and outbreak investigations. Considering the current situation in many Mediterranean countries where different lineages of WNV co-circulate with other flaviviruses, two samples containing both WNV and USUV were included in the panel to evaluate the ability of the labs to identify coinfections.

The panel for antibody detection consisted of 10 sera from WNV infected horses with different IgG and IgM antibody titers. Overall, both panels conformed a very comprehensive proficiency test that allowed accurate evaluation of the WNV and USUV diagnostic capacities of the labs.

Some limitations were observed in the generic detection of flaviviruses. On the one hand, four laboratories (23.5%) did not perform this PCR. Apparently, in these labs real-time PCR has completely replaced conventional PCR and they even lack the necessary equipment to carry out the electrophoresis. This fact can have a negative impact in their diagnostic capacities for certain pathogens. In this case, they were unable to identify the presence of a non-WNV/USUV flavivirus in sample W9 (JEV). On the other hand, out of the 13 labs that performed the pan-flavivirus conventional RT-PCR, only 46.1% reported correct results in the whole panel, identifying the eight samples containing WNV,

USUV or JEV. Most of the mistakes were false negative results, particularly in the three samples with lower viral load, for which weak positive bands were expected.

The performance of the labs with the triplex RRT-PCR was highly satisfactory in general terms, with the exception of lab #1 that reported incorrect results in all the samples. It seems that an error in the numbering of the tubes or during the transcription of the results could occur, as the reported result for each sample corresponded to the expected outcome of the preceding sample. Excluding this lab, the overall results for the specific detection of WNV L1 were excellent. The positive samples were identified by all labs except one false negative result in one lab, which represents an overall detection rate of 97.9%, much higher than that reported in previous EQAs organized by ENIVD and EVD LabNet networks in human virology labs [27,30]. With regard to the diagnosis of L2 WNV infections, these previous interlaboratory assays had evidenced important limitations in the participant labs. In the EQA organized in 2006 only 46.6% of the labs were able to detect L2 positive samples [23]. Although this percentage increased during the second ENIVD EQA in 2011, one third of the labs still failed to identify WNV L2 [24]. In the last EQA organized in 2017, the human labs had considerably improved the detection of WNV L2 but some false negative results were still reported [27]. In the present study, all the labs (except lab #1) successfully detected the presence of L2 in the positive samples of the panel. In the current epidemiological context, where L2 is already present in many Euro-Mediterranean countries and will probably expand to new territories [31,32], these results are highly relevant to ensure a timely detection of L2 circulation in animal populations of the involved countries.

As regards USUV detection, before this EQA, only two ring trials for human labs had included one USUV sample in the diagnostic panels [27,28]. Interestingly, in the EQA organized for blood transfusion centers in Italy in 2011, the USUV sample was misidentified as WNV positive by all the labs, indicating the presence of cross-reactivity in the two automated nucleic acid assays used [28]. Cross-reactions between both flaviviruses have been evidenced in several studies where automated commercial PCR kits were used to test human blood [33,34]. However, in the present EQA, USUV positive sample was reported as WNV negative by all the labs. This, together with the fact that the JEV sample was negative for the three viruses in all the labs, confirms the high specificity of the applied triplex PCR assay.

In the more recent EQA organized by EVD LabNet, the USUV sample was correctly detected by a small percentage of labs, even in countries with demonstrated USUV circulation, revealing a clear need for technical improvement in USUV diagnosis in participating labs [27]. In our EQA, 75% of the labs were able to identify the three positive samples and 81.2% detected two of them. Even so, two labs (11.7% of the total) could not detect the virus using the recommended PCR protocol due to the lack of the required fluorescent channel (Cy5) in their real-time thermocyclers.

The results of the generic antibody detection exercise were excellent, with 100% correct results in all labs except for one negative sample that one lab reported as doubtful. One limitation of this EQA is the lack of sera with antibodies directed against other flaviviruses, especially considering the high degree of cross-reactions that occur in serological assays [16]. Sera from USUV or JEV infected animals in sufficient amount to prepare an EQA panel are difficult to obtain unless they originate from experimentally infected horses. Unfortunately, we could not have access to this type of samples and, therefore, the panel was restricted to WNV positive and negative sera. As a result, we can only evaluate the capacity of the labs to successfully identify WNV antibodies, but we cannot assess the potential interference of cross-reactivity with other flaviviruses on the performance of the ELISA kits.

More difficulties were observed with the specific detection of IgM antibodies. This was most probably due to the fact that two of the 4 IgM positive sera had low IgM antibody titers with OD values close to the reference threshold. In fact, most of the incorrect results were false negatives in these two sera, while no false positive results were reported. The three labs that used an alternative kit (IDVet IgM capture ELISA) to evaluate the presence of IgM WNV antibodies were only able to detect the sample with the highest IgM titer. These results are in agreement with the data derived from the serology EQA organized by ANSES in 2013, where the INgezim WNV IgM ELISA displayed higher analytical sensitivity than the IDVet IgM capture ELISA for L1 and L2 WNV infected horses [29].

At the end of the exercise, each participant laboratory received an individual report with the analysis of their results, possible reasons for the observed deviations and recommendations to improve their competence.

The unique characteristics of this EQA, where all the participant labs used the same protocols and reagents, enables a comprehensive evaluation of the selected diagnostic methods under "controlled" conditions. In this way, we could verify that the triplex RRT-PCR protocol [15] is a reliable method for accurate detection and differentiation of WNV L1, L2 and USUV in animal samples of different type and origin. This protocol was easily transferred to all the labs that, in most cases, were able to correctly apply the assay and interpret the results. In the current epidemiological context, this method can be a very useful tool for clinical diagnostic and epidemiological surveillance of WNV and USUV. With respect to the pan-flavivirus conventional RT-PCR, more variability of results was observed, especially in samples with low viral loads. Three false positive results were also reported that could be due to the presence of unspecific bands or cross-contamination. However, this broad-range flavivirus assay is a good first-line tool for rapid flavivirus detection and a useful complement to species specific assays. Moreover, this technique allows further genome sequencing to identify the involved virus and perform phylogenetic analysis [35].

In the case of the antibody detection exercise, our results confirm the optimal performance of the Ingenasa competitive ELISA (INgezim West Nile Compac), with high reproducibility values, as all the labs reported concordant results. This commercial kit is widely used for WNV surveillance in birds and horses due to its excellent sensitivity and specificity values [16,29]. The recommended IgM ELISA (INgezim WNV IgM ELISA) also displayed good results and was able to identify positive samples that were not detected with other commercial kits.

This exercise offers a good overview of the WNV and USUV diagnostic capacities of veterinary labs in 17 EU-neighboring countries. Based on the obtained results, most of the participant labs have the necessary infrastructure and expertise to correctly perform the molecular and serological diagnosis of WNV. The training strategy developed during the MediLabSecure project, with two workshops (molecular and serological diagnosis) followed by this EQA, was beneficial in improving the capacities of the labs.

## 4. Materials and Methods

### 4.1. *Call for Participation*

An invitation letter was sent by the coordinating team of the MediLabSecure animal virology network (INIA-CISA, Madrid, Spain) to the beneficiary veterinary laboratories ($n$ = 18). Seventeen laboratories accepted to participate (94.4%). The participation was free of charge and entailed the publication of comparative results in an anonymous manner.

### 4.2. *Preparation of EQA Panel*

#### 4.2.1. Samples for Virus Genome Detection

For the molecular diagnosis of WNV, each participant received a coded panel of 10 samples, as shown in Table 1.

Four viral strains were used for the preparation of the panel: SP07 strain (WNV L1), isolated from a golden eagle in Spain in 2007 [36]; AUS08 strain (WNV L2), isolated from a goshawk in Austria in 2008 [37], USU11 (Usutu virus) isolated from a blackbird in Italy in 2011 (GenBank number KX816649) [38] and the Nakayama strain (Japanese encephalitis virus-JEV). All the viral stocks were inactivated using ß-propiolactone. Absence of residual infectivity was confirmed after three consecutive passages in Vero cells by absence of cyto-phatic effect and by RRT-PCR analysis.

Several dilutions of inactivated viral stocks were spiked in different matrices (serum, blood, liver, heart or kidney) from healthy non-infected birds and horses to prepare the positive samples. The negative samples consisted of brain and heart homogenates from healthy birds and horses. Nucleic acid extraction was performed from 200 μL of sample using the QIAamp® Cador Pathogen Mini Kit (QIAGEN), following the manufacturer's instructions. In the final step, RNA was eluted in 50 μL of nuclease-free water. All samples were tested twice with two validated and widely used PCR techniques that we selected as recommended methods: a conventional RT-PCR for pan-flavivirus detection [35] and a RRT-PCR for simultaneous WNV and USUV detection [15].

For the conventional RT-PCR, mix was prepared in a final volume of 25 μL per sample containing 2 μL of RNA template, 0.6 μM of each primer (cFD2 and MAMD), RT-PCR enzyme mix and RT-PCR buffer of the commercial SuperScript® III One-Step RT-PCR System with Platinum® Taq DNA polymerase (Life Technologies, Thermo Fisher Scientific). All reactions were carried out using the following thermal profile: reverse transcription at 55 °C for 30 min, initial PCR activation step at 94 °C for 2 min, followed by 40 cycles of 30 s at 94 °C, 30 s at 55 °C, and 30 s at 68 °C and a final extension step of 5 min at 68 °C. Amplified products were analyzed by 2% agarose gel electrophoresis. Positive samples should give a specific band of the same size as the positive control (252 bp).

The RRT-PCR was performed using the primers, probes and the thermal profile described by del Amo et al. [15]. Samples with Ct > 40 were considered negative.

According to the obtained bands in the conventional RT-PCR and the Ct values in the triplex RRT-PCR, a collection of 10 samples was finally selected (Tables 1 and 2). The samples were aliquoted (1 mL) and each vial was lyophilized and stored at 4 °C until delivery to the participant laboratories.

Prior to delivery, the lyophilized panel was resuspended in DNAse-free water and was fully analyzed to verify the integrity of the samples and the reproducibility of the results after lyophilization. Triplicates of each lyophilized sample were analyzed by 3 technicians at INIA-CISA using the mentioned techniques. For the RRT-PCR, the reference Ct value was established as the mean of the nine repetitions (Table 2).

Two positive controls were delivered with the panel: (1) a triplex positive extraction control consisting of cell culture medium spiked with a mix of inactivated WNV L1, L2 and USUV strains to obtain, after a 1/10 dilution, an expected Ct value of 32 ± 2 for each virus (this sample was lyophilized and stored at 4 °C until delivery) and (2) a triplex positive reaction control consisting of a mix of WNV L1, L2 and USUV RNAs with an expected Ct value of 32 ± 2 for each virus (this sample was stored at −80 °C until delivery).

#### 4.2.2. Samples for Antibody Detection

For the serological diagnosis of WNV, each participant received a panel of 10 samples (six positive and four negative) (Table 3). The positive samples included sera obtained from naturally infected horses in Southern Spain and one positive reference serum from the EU Reference Laboratory for equine diseases. Four of these samples were IgM positive, obtained from recently infected horses (Table 3). The negative samples consisted of one serum from a non-infected horse and a commercial negative horse serum (Biowhittaker). All samples were inactivated by heating at 56 °C for 45 min. Each sample was aliquoted (130 μL) and stored at −20 °C until delivery.

### *4.3. EQA Details*

For the molecular detection, two assays were proposed. First, for generic detection of flaviviruses we recommended the hemi-nested conventional RT-PCR described by Scaramozzino et al. [35] with some modifications as described in the previous section. This RT-PCR targets the highly conserved NS5 region and it is a useful first-line molecular screening test for an unknown flavivirus. However, definitive flavivirus determination requires post-amplification identification techniques (e.g., genome sequencing) or the application of RT-PCR techniques for species-specific detection of flaviviruses. For this reason, the second recommended assay was a triplex RRT-PCR that enables simultaneous detection and differentiation of WNV L1, L2 and USUV [15]. This is based on different sets of primers and fluorogenic probes specific to each virus that are labelled with selective, non-overlapping fluorogen-quencher pairs (FAM for WNV L1, VIC for WNV L2 and Cy5 for USUV). This multiplex RRT-PCR is very sensitive and specific and has been widely validated with experimental and field samples [15,39]. Taking into account the epidemiological situation of the concerned region, where different WNV lineages co-circulate with other related flaviviruses, and especially USUV, this diagnostic approach was considered highly beneficial for surveillance and diagnostic studies.

The labs reported the use of eight different real-time thermocyclers: Rotor-Gene 3000 (5 labs), Applied Biosystems 7500 (5 labs), Applied Biosystems 7300 (2 labs), Aria Mx Agilent (1 lab), Bioer Gene Max (1 lab), Abi QuantStudio (1 lab), StepOnePlus (1 lab) and Stratagene Mx3005 (1 lab).

For serological diagnosis of WNV infection, two commercially available ELISA tests were recommended. The first was the INgezim West Nile Compac ELISA kit (Ingenasa) that allows the detection of anti-E domain III antibodies [40]. It is a multispecies ELISA that only requires 10 µL of sample which is very advantageous for the analysis of sera from small birds. Several studies have proved that this assay is more specific than other commercial ELISAs that, although designed to identify WNV antibodies, also detect cross-reacting antibodies directed against other flaviviruses and especially USUV [16,29]. Moreover, the INgezim West Nile Compac ELISA detects both IgG and IgM antibodies and, based on the results of previous EQAs, it seems that this kit enables more efficient detection of recently-infected animals than other commercial kits [29]. To specifically identify recent (acute) infection in horses, we selected an IgM antibody capture ELISA (MAC-ELISA) from the same company, the Ingezim WNV IgM ELISA. This assay was recommended based on our own comparative studies and on the results of an EQA organized in 2013 where this kit demonstrated the highest analytical sensitivity in horses experimentally infected with WNV L1 and L2 [29].

Detailed standard operating procedures for the recommended assays were distributed and all the reagents and kits were provided to each laboratory, including the mentioned ELISA kits, the extraction kit (QIAamp® Cador Pathogen Mini Kit, QIAGEN), the RT-PCR kit (SuperScript® III One-Step RT-PCR System, Invitrogen) and primers for generic flavivirus detection, the RRT-PCR kit (QuantiTect Probe RT-PCR kit, QIAGEN), and primers and probes for the triplex RRT-PCR. As explained earlier, extraction and reaction positive controls were also delivered to all the labs.

For the molecular panel, specific instructions were provided to reconstitute the lyophilized samples as well as to prepare the positive extraction control. For the serological panel, the laboratories were asked to analyze the panel samples using the recommended kits following manufacturer's instructions. Additionally, the labs were encouraged to analyze both panels using alternative methods (other protocols that may be established in the labs) and report the results together with those derived from the recommended methods.

The extraction kit, the lyophilized samples, the positive extraction control and the ELISA kits were shipped at room temperature. The panel of sera, the RT-PCR kits, the positive reaction control and the primers and probes were shipped in dry ice. A number code was assigned to each laboratory to ensure a blind analysis of the results.

**Author Contributions:** Conceptualization: J.F.-P., M.Á.J.-C., E.P.-R., C.C.-G., F.L.; Formal analysis: E.P.-R., C.C.-G., F.L.; Funding acquisition: M.Á.J.-C.; Investigation: E.P.-R., C.C.-G., F.L., J.F.-P., M.Á.J.-C., A.V., L.V., N.T., K.S., S.S., A.O., A.K., K.K., I.K.-H., N.M.H., J.E.H., H.D., M.S.B., B.A., N.A. Methodology: E.P.-R., C.C.-G., F.L., J.F.-P., M.A.J.-C. Project administration: M.Á.J.-C., E.P.-R.; Supervision: J.F.-P., M.Á.J.-C.; Writing—original draft: E.P.-R.; Writing—review & editing: E.P.-R., C.C.-G., F.L., J.F.-P., M.Á.J.-C. All authors have read and agreed to the published version of the manuscript.

**Funding:** MediLabSecure project is supported by the European Commission (EU DG DEVCO: IFS/21010/23/-194 and 2018/402-247). www.medilabsecure.com.

**Acknowledgments:** We thank María del Carmen Barbero, Pilar Aguilera, Ana María Robles and Amalia Villalba for excellent technical assistance during preparation and testing of the panels. We are grateful to Ana Moreno (IZSLER, Brescia), Norbert Nowotny (University of Vienna) and Ana Vázquez (ISCIII, Spain) for kindly providing the USUV, WNV L2 and Nakayama JEV strains that were used to prepare the molecular panel. We thank Jordi Figuerola and Ramón Soriguer (EBD-CSIC, Spain) for providing the sera from naturally infected horses and to Sylvie Lecollinet (ANSES, France) for providing the reference serum. We are also grateful to Lyudmila Maruschak and Elena Coada for participating in the analysis of the panels in Ukraine and Moldova.

**Conflicts of Interest:** The authors declare no conflict of interest. The contents of this article are the sole responsibility of the authors and do not necessarily reflect the views of the European Union.

**Ethical Statement:** The tissue samples and sera used for the preparation of the EQA panels were selected from the biobank of INIA-CISA (Madrid, Spain). None of them were specifically collected for this study. All the samples used to prepare the molecular panel originated from animal studies that were authorized by INIA Animal Experimentation Ethics Committee according to European Directive 2010/63/EU (Spanish Royal Decree 53/2013). The sera used in the serological panel were obtained from naturally infected horses in Doñana National park (Southern Spain) that were sampled by expert veterinarians in the framework of a WNV sero-surveillance research project. All procedures to obtain these samples were approved by CSIC Ethics Committee.

## References

1. Colpitts, T.M.; Conway, M.J.; Montgomery, R.R.; Fikrig, E. West Nile virus: Biology, transmission, and human infection. *Clin. Microbiol. Rev.* **2012**, *25*, 635–648. [CrossRef] [PubMed]
2. McLean, R.G.; Ubico, S.R.; Docherty, D.E.; Hansen, W.R.; Sileo, L.; McNamara, T.S. West Nile virus transmission and ecology in birds. *Ann. N. Y. Acad. Sci.* **2001**, *951*, 54–57. [CrossRef] [PubMed]
3. Chancey, C.; Grinev, A.; Volkova, E.; Rios, M. The global ecology and epidemiology of West Nile virus. *BioMed Res. Int.* **2015**, *2015*, 1–20. [CrossRef] [PubMed]
4. Castillo-Olivares, J.; Mansfield, K.L.; Phipps, L.P.; Johnson, N.; Tearle, J.; Fooks, A.R. Antibody response in horses following experimental infection with West Nile virus lineages 1 and 2. *Transbound. Emerg. Dis.* **2011**, *58*, 206–212. [CrossRef] [PubMed]
5. Pérez-Ramírez, E.; Llorente, F.; Jiménez-Clavero, M.A. Experimental infections of wild birds with West Nile virus. *Viruses* **2014**, *6*, 752–781. [CrossRef] [PubMed]
6. Beck, C.; Jiménez-Clavero, M.A.; Leblond, A.; Durand, B.; Nowotny, N.; Leparc-Goffart, I.; Zientara, S.; Jourdain, E.; Lecollinet, S. Flaviviruses in Europe: Complex circulation patterns and their consequences for the diagnosis and control of West Nile disease. *Int. J. Environ. Res. Public Health* **2013**, *10*, 6049–6083. [CrossRef]
7. Gossner, C.M.; Marrama, L.; Carson, M.; Allerberger, F.; Calistri, P.; Dilaveris, D.; Lecollinet, S.; Morgan, D.; Nowotny, N.; Paty, M.-C.; et al. West Nile virus surveillance in Europe: Moving towards an integrated animal-human-vector approach. *Eurosurveillance* **2017**, *22*, 30526. [CrossRef]
8. ECDC. Epidemiological Update: West Nile virus Transmission Season in Europe. 2018. Available online: https://ecdc.europa.eu/en/news-events/epidemiological-update-west-nile-virus-transmission-season-europe-2018 (accessed on 12 October 2020).
9. Clé, M.; Beck, C.; Salinas, S.; Lecollinet, S.; Gutiérrez, S.; Van de Perre, P.; Baldet, T.; Foulongne, V.; Simonin, Y. Usutu virus: A new threat? *Epidemiol. Infect.* **2019**, *147*, e232. [CrossRef]
10. Weissenböck, H.; Kolodziejek, J.; Url, A.; Lussy, H.; Rebel-Bauder, B.; Nowotny, N. Emergence of Usutu virus, an African mosquito-borne flavivirus of the Japanese encephalitis virus group, Central Europe. *Emerg. Infect. Dis.* **2002**, *8*, 652–656. [CrossRef]
11. Nikolay, B. A review of West Nile and Usutu virus co-circulation in Europe: How much do transmission cycles overlap? *Trans. R. Soc. Trop. Med. Hyg.* **2015**, *109*, 609–618. [CrossRef]
12. Durand, B.; Haskouri, H.; Lowenski, S.; Vachiery, N.; Beck, C.; Lecollinet, S. Seroprevalence of West Nile and Usutu viruses in military working horses and dogs, Morocco, 2012: Dog as an alternative WNV sentinel species? *Epidemiol. Infect.* **2016**, *144*, 1857–1864. [CrossRef] [PubMed]
13. Ben Hassine, T.; De Massis, F.; Calistri, P.; Savini, G.; BelHaj Mohamed, B.; Ranen, A.; Di Gennaro, A.; Sghaier, S.; Hammami, S. First detection of co-circulation of West Nile and Usutu viruses in equids in the South-west of Tunisia. *Transbound. Emerg. Dis.* **2014**, *61*, 385–389. [CrossRef] [PubMed]

14. Kemenesi, G.; Buzás, D.; Zana, B.; Kurucz, K.; Krtinic, B.; Kepner, A.; Földes, F.; Jakab, F. First genetic characterization of Usutu virus from Culex pipiens mosquitoes Serbia, 2014. *Infect. Genet. Evol.* **2018**, *63*, 58–61. [CrossRef] [PubMed]
15. del Amo, J.; Sotelo, E.; Fernández-Pinero, J.; Gallardo, C.; Llorente, F.; Agüero, M.; Jiménez-Clavero, M.A. A novel quantitative multiplex real-time RT-PCR for the simultaneous detection and differentiation of West Nile virus lineages 1 and 2, and of Usutu virus. *J. Virol. Methods* **2013**, *189*, 321–327. [CrossRef] [PubMed]
16. Llorente, F.; García-Irazábal, A.; Pérez-Ramírez, E.; Cano-Gómez, C.; Sarasa, M.; Vázquez, A.; Jiménez-Clavero, M.Á. Influence of flavivirus co-circulation in serological diagnostics and surveillance: A model of study using West Nile, Usutu and Bagaza viruses. *Transbound. Emerg. Dis.* **2019**, *66*, 2100–2106. [CrossRef] [PubMed]
17. Riccardo, F.; Monaco, F.; Bella, A.; Savini, G.; Russo, F.; Cagarelli, R.; Dottori, M.; Rizzo, C.; Venturi, G.; Di Luca, M.; et al. An early start of West Nile virus seasonal transmission: The added value of One Heath surveillance in detecting early circulation and triggering timely response in Italy, June to July 2018. *Eurosurveillance* **2018**, *23*, 1800427. [CrossRef]
18. Escadafal, C.; Gaayeb, L.; Riccardo, F.; Pérez-Ramírez, E.; Picard, M.; Dente, M.G.; Fernández-Pinero, J.; Manuguerra, J.-C.; Jiménez-Clavero, M.Á.; Declich, S.; et al. Risk of Zika virus transmission in the Euro-Mediterranean area and the added value of building preparedness to arboviral threats from a One Health perspective. *BMC Public Health* **2016**, *16*, 1219. [CrossRef]
19. Dente, M.; Riccardo, F.; Nacca, G.; Ranghiasci, A.; Escadafal, C.; Gaayeb, L.; Jiménez-Clavero, M.Á.; Manuguerra, J.-C.; Picard, M.; Fernández-Pinero, J.; et al. Strengthening preparedness for arbovirus infections in Mediterranean and Black Sea countries: A conceptual framework to assess integrated surveillance in the context of the One Health strategy. *Int. J. Environ. Res. Public Health* **2018**, *15*, 489. [CrossRef]
20. Pérez-Ramírez, E.; Cano-Gómez, C.; Llorente, F.; Adzic, B.; Al Ameer, M.; Djadjovski, I.; El Hage, J.; El Mellouli, F.; Goletic, T.; Hovsepyan, H.; et al. External quality assessment of Rift Valley fever diagnosis in 17 veterinary laboratories of the Mediterranean and Black Sea regions. *PLoS ONE* **2020**, *15*, e0239478. [CrossRef]
21. Eiden, M.; Vina-Rodríguez, A.; Hoffmann, B.; Ziegler, U.; Groschup, M.H. Two new real-time quantitative reverse transcription polymerase chain reaction assays with unique target sites for the specific and sensitive detection of lineages 1 and 2 West Nile virus strains. *J. Vet. Diagn. Investig.* **2010**, *22*, 748–753. [CrossRef]
22. Weissenböck, H.; Bakonyi, T.; Chvala, S.; Nowotny, N. Experimental Usutu virus infection of suckling mice causes neuronal and glial cell apoptosis and demyelination. *Acta Neuropathol.* **2004**, *108*, 453–460. [CrossRef] [PubMed]
23. Niedrig, M.; Linke, S.; Zeller, H.; Drosten, C. First international proficiency study on West Nile virus molecular detection. *Clin. Chem.* **2006**, *52*, 1851–1854. [CrossRef] [PubMed]
24. Linke, S.; MacKay, W.G.; Scott, C.; Wallace, P.; Niedrig, M. Second external quality assessment of the molecular diagnostic of West Nile virus: Are there improvements towards the detection of WNV? *J. Clin. Virol.* **2011**, *52*, 257–260. [CrossRef] [PubMed]
25. Niedrig, M.; Donoso Mantke, O.; Altmann, D.; Zeller, H. First international diagnostic accuracy study for the serological detection of West Nile virus infection. *BMC Infect. Dis.* **2007**, *7*, 72. [CrossRef] [PubMed]
26. Sanchini, A.; Donoso-Mantke, O.; Papa, A.; Sambri, V.; Teichmann, A.; Niedrig, M. Second international diagnostic accuracy study for the serological detection of West Nile virus infection. *PLoS Negl. Trop. Dis.* **2013**, *7*, e2184. [CrossRef] [PubMed]
27. Reusken, C.; Baronti, C.; Mögling, R.; Papa, A.; Leitmeyer, K.; Charrel, R.N. Toscana, West Nile, Usutu and tick-borne encephalitis viruses: External quality assessment for molecular detection of emerging neurotropic viruses in Europe, 2017. *Eurosurveillance* **2019**, *24*, 1900051. [CrossRef]
28. Pisani, G.; Pupella, S.; Cristiano, K.; Marino, F.; Simeoni, M.; Luciani, F.; Scuderi, G.; Sambri, V.; Rossini, G.; Gaibani, P.; et al. Detection of West Nile virus RNA (lineages 1 and 2) in an external quality assessment programme for laboratories screening blood and blood components for West Nile virus by nucleic acid amplification testing. *Blood Transfus.* **2012**, *10*, 515–520.
29. Beck, C.; Lowenski, S.; Durand, B.; Bahuon, C.; Zientara, S.; Lecollinet, S. Improved reliability of serological tools for the diagnosis of West Nile fever in horses within Europe. *PLoS Negl. Trop. Dis.* **2017**, *11*, e0005936. [CrossRef]

30. Sambri, V.; Capobianchi, M.R.; Cavrini, F.; Charrel, R.; Donoso-Mantke, O.; Escadafal, C.; Franco, L.; Gaibani, P.; Gould, E.A.; Niedrig, M.; et al. Diagnosis of West Nile virus human infections: Overview and proposal of diagnostic protocols considering the results of external quality assessment studies. *Viruses* **2013**, *5*, 2329–2348. [CrossRef]
31. HernÃ¡ndez-Triana, L.M.; Jeffries, C.L.; Mansfield, K.L.; Carnell, G.; Fooks, A.R.; Johnson, N. Emergence of West Nile virus lineage 2 in Europe: A review on the introduction and spread of a mosquito-borne disease. *Front. Public Health* **2014**, *2*, 271. [CrossRef]
32. Napp, S.; Petrić, D.; Busquets, N. West Nile virus and other mosquito-borne viruses present in Eastern Europe. *Pathog. Glob. Health* **2018**, *112*, 233–248. [CrossRef] [PubMed]
33. Cadar, D.; Maier, P.; Müller, S.; Kress, J.; Chudy, M.; Bialonski, A.; Schlaphof, A.; Jansen, S.; Jöst, H.; Tannich, E.; et al. Blood donor screening for West Nile virus (WNV) revealed acute Usutu virus (USUV) infection, Germany, September 2016. *Eurosurveillance* **2017**, *22*, 30501. [CrossRef] [PubMed]
34. Carletti, F.; Colavita, F.; Rovida, F.; Percivalle, E.; Baldanti, F.; Ricci, I.; De Liberato, C.; Rosone, F.; Messina, F.; Lalle, E.; et al. Expanding Usutu virus circulation in Italy: Detection in the Lazio region, central Italy, 2017 to 2018. *Eurosurveillance* **2019**, *24*, 1800649. [CrossRef] [PubMed]
35. Scaramozzino, N.; Crance, J.-M.; Jouan, A.; DeBriel, D.; Stoll, F.; Garin, D. Comparison of flavivirus universal primer pairs and development of a rapid, highly sensitive heminested reverse transcription-PCR assay for detection of flaviviruses targeted to a conserved region of the NS5 gene sequences. *J. Clin. Microbiol.* **2001**, *39*, 1922–1927. [CrossRef]
36. Jiménez-Clavero, M.A.; Sotelo, E.; Fernández-Pinero, J.; Llorente, F.; Blanco, J.M.; Rodriguez-Ramos, J.; Pérez-Ramírez, E.; Höfle, U. West Nile virus in Golden eagles, Spain, 2007. *Emerg. Infect. Dis.* **2008**, *14*, 1489–1491. [CrossRef]
37. Barzon, L.; Pacenti, M.; Franchin, E.; Lavezzo, E.; Masi, G.; Squarzon, L.; Pagni, S.; Toppo, S.; Russo, F.; Cattai, M.; et al. Whole genome sequencing and phylogenetic analysis of West Nile virus lineage 1 and lineage 2 from human cases of infection, Italy, August 2013. *Eurosurveillance* **2013**, *18*, 20591. [CrossRef]
38. Calzolari, M.; Chiapponi, C.; Bonilauri, P.; Lelli, D.; Baioni, L.; Barbieri, I.; Lavazza, A.; Pongolini, S.; Dottori, M.; Moreno, A. Co-circulation of two Usutu virus strains in Northern Italy between 2009 and 2014. *Infect. Genet. Evol.* **2017**, *51*, 255–262. [CrossRef]
39. Pérez-Ramírez, E.; Llorente, F.; del Amo, J.; Nowotny, N.; Jiménez-Clavero, M.Á. Susceptibility and role as competent host of the red-legged partridge after infection with lineage 1 and 2 West Nile virus isolates of Mediterranean and Central European origin. *Vet. Microbiol.* **2018**, *222*, 39–45. [CrossRef]
40. Sotelo, E.; Llorente, F.; Rebollo, B.; Camuñas, A.; Venteo, A.; Gallardo, C.; Lubisi, A.; Rodríguez, M.J.; Sanz, A.J.; Figuerola, J.; et al. Development and evaluation of a new epitope-blocking ELISA for universal detection of antibodies to West Nile virus. *J. Virol. Methods* **2011**, *174*, 35–41. [CrossRef]


**Publisher's Note:** MDPI stays neutral with regard to jurisdictional claims in published maps and institutional affiliations.

pathogens

MDPI

*Review*

# Animal and Human Vaccines against West Nile Virus

**Juan-Carlos Saiz**

Department of Biotechnology, Instituto Nacional de Investigación y Tecnología Agraria y Alimentaria (INIA), 28040 Madrid, Spain; jcsaiz@inia.es; Tel.: +34-913471497

Received: 18 November 2020; Accepted: 17 December 2020; Published: 21 December 2020

**Abstract:** West Nile virus (WNV) is a widely distributed enveloped flavivirus transmitted by mosquitoes, which main hosts are birds. The virus sporadically infects equids and humans with serious economic and health consequences, as infected individuals can develop a severe neuroinvasive disease that can even lead to death. Nowadays, no WNV-specific therapy is available and vaccines are only licensed for use in horses but not for humans. While several methodologies for WNV vaccine development have been successfully applied and have contributed to significantly reducing its incidence in horses in the US, none have progressed to phase III clinical trials in humans. This review addresses the status of WNV vaccines for horses, birds, and humans, summarizing and discussing the challenges they face for their clinical advance and their introduction to the market.

**Keywords:** flavivirus; West Nile virus; human vaccines; animal vaccines

---

## 1. Introduction

Pathogen inoculation as a preventive measure against infectious diseases (namely smallpox) had been used for centuries in China and other parts of the world. However, and although other researchers had applied similar principles before, the credit for the first effective vaccine goes to Edward Jenner who, in 1798, published the first evidence that supported its efficacy in preventing smallpox [1]. Since then, vaccine development and implementation have been continuously growing until became a breakthrough for animal and human health. Nevertheless, and even though vaccination campaigns are saving millions of lives yearly, nowadays, significant challenges still remain in the vaccinology field, as is exemplified by the devastating current SARS-CoV-2 pandemic. For example, the need: (i) for new technologies and adjuvants that allow faster development of more effective, stable, and low-cost vaccines; (ii) that the population is duly informed about the benefits of vaccines as important for their well-being and health, counteracting the increase in the dissemination of opinions of anti-vaccine groups and deniers observed in recent years with data and arguments easily understood by the population to, in this way, obtain their informed consent; (iii) to maintain investment in this field, regardless of the perception of its relevance for the global health of society as a whole and pharmaceutical concerns for the return on investment, thus avoiding situations such as those experienced in the case of Ebola vaccines, for which several candidates have demonstrated their efficacy in animal models, but none has been authorized, which will hamper its control against possible new outbreaks; and (iv) to establish a global vaccine development fund since the current business model only prioritizes those with great market potential [2].

Among the main targets of viral vaccination are (re) emerging pathogens that have caused recent epidemics/pandemics around the world, such as avian influenza, Ebola, SARS, MERS, Chikungunya, Zika, Dengue, and West Nile viruses, as well as those that may pose a threat in the future. This review describes the status of West Nile virus (WNV) vaccines for horses, birds, and humans, summarizing and discussing the challenges they face in their clinical advance and introduction to the market.

## 2. West Nile Virus

WNV is an enveloped flavivirus (family *Flaviviridae*) transmitted by mosquitoes, mainly of the *Culex pipiens* L. complex, whose natural hosts are birds [3]. The virus occasionally infects other vertebrates, mainly equids and humans, which are accidental "dead-end" hosts because viremias achieved in mammals are usually inadequate to maintain the virus cycle, being not enough high to infect a naïve mosquito while feeding on them [4]. Nevertheless, WNV infections in humans and horses have great economic and health repercussions. Although most WNV infections in humans are asymptomatic, around 20% may cause West Nile fever and less than one percent West Nile neuroinvasive disease, which may result in febrile illness, meningitis, encephalitis, flaccid paralysis, and even death, which can occur in around 10% of severe cases [5,6]. In fact, WNV is the arthropod-borne human pathogenic virus with the largest distribution and one of the major causes of human viral encephalitis worldwide [5,6].

WNV is classified into several lineages that do not consistently correlate with its geographical distribution, but only lineages 1 and 2 have been involved in human outbreaks [7]. Early reports suggested that both lineages had differences in pathogenicity, virulence, viremia, the clinical course of infection, and mortality. This initial hypothesis was based mainly on the lack of clinical incidence of WNV in Africa, where only mild diseases have been reported and no deaths have been documented in humans, and where lineage 2 was restricted until it colonized Europe [8]. However, later data in humans and in naturally or experimentally infected animals dismantled this hypothesis. Thus, a study of 644 Greek individuals, which provided a suitable blood sample and lived in an area suffering a WNV lineage 2 epidemic, showed that 5.8% were seropositive for WNV-specific IgG and approximately 18% of them presented clinical manifestations of WNV disease, figures similar to those of patients infected with lineage 1 [9]. Likewise, falcons and magpies experimentally infected with strains of lineage 1 or 2 showed similar mortality rates [10,11]. Even more, studies performed in vaccinated animals showed a high degree of cross-protection between both lineages. Mice immunized with the inactivated Duravaxyn WNV vaccine or with an experimental RSP (Recombinant Subviral Particle) candidate, both based on lineage 1 strains, were protected against exposure to lineage 2 strains [12,13]. Similarly, horses experimentally vaccinated with the ALVAC (®)-WNV vaccine [14], or vaccinated under field conditions with the inactivated Equip vaccine [15], both based on lineage 1 strains, were also protected against challenge with heterologous strains of lineage 2.

WNV genome is a single-stranded RNA molecule of positive polarity that encodes three structural (E, prM/M, and C) and seven non-structural (NS1, 2A, 2B, 3, 4A, 4B, and 5) proteins [3]. Among the structural proteins, the E glycoprotein, which is involved in receptor binding, viral entry, and membrane fusion, is the most immunogenic one [16]. This protein has three domains (DI, DII, and DIII), being DIII an immunoglobulin-like structure that contains multiple epitopes recognized by neutralizing antibodies [3]. In fact, several flaviviruses share common epitopes recognized by cross-reactive neutralizing antibodies [17], which may have consequences for the implementation of vaccines, mainly in regions where several of them co-circulate.

## 3. Vaccines

Despite the great efforts invested in recent years in the development of prophylactic measures against this pathogen, there is currently no specific drug or therapy licensed for its treatment [18,19]. However, several candidate vaccines have been successfully developed, some of which have been licensed for use in horses but, despite the fact that no adverse events or safety concerns have been reported in the few clinical trials conducted, none has been authorized for humans. For the development of the WNV vaccine, all possible approaches available have been tested, from purified inactivated and live attenuated viruses, to candidates based on nucleic acids (DNA or RNA), virus-like particles, subunit elements, and recombinant viruses.

### 3.1. Animal Vaccines

As commented before, equids are sporadically infected by WNV and, although in most cases they remain asymptomatic, around 20% can develop clinical signs that use to be more severe than in humans and have important health and economic consequences [20]. When present, signs can range from fever to infection and inflammation of the nervous system, and can cause ataxia, hind and forelimb weakness, quadriplegia, paresis, seizures, chewing and paralysis of the tongue, depression, and ophthalmologic manifestations. [20,21]. Therefore, great efforts have been made to develop and implement equine vaccines. Four of the six licensed vaccines are currently on the market for use in horses. The WN-Innovator, with a classic inactivated whole virion-based approach, was the first to be developed and was licensed by the USDA in 2003 [22]. Live attenuated recombinant viruses have also been used (either based on canary poxvirus or yellow fever virus), as well as a plasmid DNA vaccine, which was the first licensed by the USDA [23], although it was subsequently withdrawn from the market by the manufacturers. All these vaccines are shown to be protective and their use has contributed greatly to reducing the incidence of the disease in horses in the US [23,24]. However, despite their proven efficacy, these vaccines still exhibit some limitations, as the need for repeated administrations to get a solid initial immunization, and the relatively short duration of the induced immunity, which makes necessary annual boosters.

Birds are the natural hosts of WNV and play a key role in the epidemiology of the virus, being many species susceptible to the infection, particularly corvids [4,25]. The disease shows up due to virus invasion of different organs: liver, spleen, kidney, heart, and mainly the central nervous system, and can lead to death within 24–48 h later [26,27]. Several commercial and experimental vaccine candidates have been assayed in wild and domestic birds, although not one has yet been authorized for use on them [27]. Overall, they induced humoral and, although less analyzed, cellular responses, and reduced disease, injury, viremia, viral shedding, and mortality associated with WNV. Furthermore, if they induce herd immunity, they could help prevent outbreaks and the spread of the virus. For example, prospective vaccination of the entire population of California condors (*Gymnogyps californianus*), an endangered species, before the arrival of WNV would have helped prevent infection and its possible extinction [28]. Likewise, vaccination also greatly reduced virus incidence in domestic geese in Israel [29]. However, the implementation of bird vaccines faces several drawbacks, as the feasibility of access to the target host, mainly for wild species, and the administration route. In any case, its availability could benefit domestic populations (farm birds, including those for hunting and restocking activities), as well as wild ones (as those housed in rehabilitation centers and wildlife reserves, and in recreational facilities, like zoos) [27].

### 3.2. Human Vaccines

As mentioned above, human WNV outbreaks can have serious health repercussions that the availability of licensed vaccines would help to minimize.

Human vaccines must be cost-effective, protective, and safe, especially for the most vulnerable populations, like the elderly and the immunosuppressed, whose numbers are increasing around the world. Ideally, vaccines should also be strongly immunogenic and long-lasting with a single dose. Furthermore, although neutralizing antibodies are currently the most reliable protective correlate for flavivirus infections, vaccines should also include determinants that stimulate a balanced T-cell response, essential for providing an effective protective response against infection [30].

Nowadays, effective licensed vaccines, either attenuated (yellow fever, Dengue, and Japanese encephalitis), or inactivated (Japanese encephalitis, tick-borne encephalitis, and Kyasanur forest disease), are available against several flaviviruses. Nevertheless, none has been licensed for human use against WNV, and none of the six vaccines assayed in humans have progressed further than to phase I/II clinical trials [31]. Even more, only two attenuated recombinant candidates expressing the WNV prM and E proteins, ChimeriVax (in a yellow fever virus backbone) and rWN/DEN4Δ30 (in a truncated Dengue virus 4 backbone), induced strong immunity after a single dose [32,33]. Noteworthy, the immunogenicity of these vaccines has been analyzed based on seroconversion and detection of neutralizing antibody, and the development of a T cell-specific response has hardly been addressed.

Therefore, for the implementation of human vaccines, several factors must still have to be analyzed in depth. Among them, the possibility that they induce disease associated with pre-existing immunity to heterologous flaviviruses should be avoided, as the phenomenon of antibody-dependent enhancement (ADE) that can arise from the binding of antibodies that cannot neutralize the virus, and can lead to increased uptake of virus into host cells through Fc receptor-mediated endocytosis in macrophages. Consequently, although its role in the pathogenesis of flavivirus infections other than Dengue remains controversial [34], ADE might lead to more severe symptoms during a secondary, heterologous infection, as it may happen in dengue virus infections [35]. Although it has been reported that DIII does not induce ADE for other flaviviruses [36,37], this hypothetical drawback could be resolved by modifying protein E with mutations in its DIII domain, or nearby, to reduce the binding of antibodies induced against other flaviviruses [38,39]. In any case, addressing and excluding this possibility is necessary during WNV vaccine development.

### *3.3. Current Challenges for Human VACCINES Implementation*

Several issues hamper the implementation and introduction of WNV vaccines into the market, such as scientific challenges, safety considerations, difficulties in setting up clinical trials, and cost-effectivity issues, among others.

As mentioned, one of the main scientific questions to solve is the induction of a complete and long-lasting protective immunity that, although seroprevalence studies and experimentation with animal models suggest that this is the case, still needs to be verified in real conditions. In addition, ideally, vaccines should also include determinants that induce a strong cellular response. Likewise, it is desirable that immunization will be induced after the administration of a single dose, and, if possible, that vaccines are marked, that is, that they are DIVA (Differentiating Infected from VAccinated individuals), which could have a considerable impact on blood donation.

Further, clinical trials are difficult to conduct because WNV outbreaks occur sporadically and often unpredictably in regions where other cross-reactive flaviviruses co-circulate. Hence, it becomes difficult to properly assess the impact of candidate vaccines and their implementation strategies.

In addition, another fact to take into account is the possibility of the imposition of new antigenic variants on a previous dominant one, as exemplified by the lineage 1 WN02 strain that has replaced the original lineage 1 NY99 strain in the USA [40]. Although WNV has to infect and replicate in different hosts, making it difficult for mutations to fixate on its genome because they can be beneficial for infecting one host (mosquitoes) but detrimental to others (birds), vaccine production must be conscious of the possibility that new antigenic variants appear and are imposed in order to minimize their possible consequences.

Ultimately, vaccines should be inexpensive, but keep in mind that, while still debated, the few studies conducted so far have concluded that current WNV vaccines are unlikely to save costs [41,42].

## 4. Conclusions

WNV remains a significant threat to humans, equids, and birds in many parts of the world, as demonstrated by the worrisome increase of cases reported in the E.U. in recent years. An accumulated number of 285 human cases and 31 deaths were reported to the European Centre for Disease Prevention and Control (ECDC) in 2020 [43]. This increase is favored, among other reasons, by the introduction or imposition of new lineages, the increase in temperature that can favor the colonization of new regions by mosquito vectors competent for their transmission, and the globalization of trade and the transport of people and animals. All this makes it necessary to have effective and inexpensive vaccines for immediate use.

As has been detailed, the technologies to produce vaccines are already available and have proved to be safe and protective in the few clinical trials conducted. In this regard, and even though clinical trials are difficult to establish, vaccination in restricted regions where outbreaks appear could help not only to combat the spread of infection but also to get a better idea of the performance of vaccines in terms of protection, the durability of immunity, etc. In any case, some issues should be improved and properly addressed to facilitate its introduction into the market. Among them are a solid demonstration of the induction of a long-lasting immunity, preferably after a single immunization, the production and validation of marked candidates, the exclusion of the possibility of inducting a putative ADE phenomenon, its safety in elderly and immunocompromised people, and the optimization of their production to reduce their costs. To solve these issues, a joint effort of the different agents involved (pharmaceutical companies, governments, international institutions, and non-governmental organizations) to establish a common fund for its development, clinical trials implementation, and the establishment of adequate vaccination strategies would be desirable.

**Funding:** This research was funded by Instituto Nacional de Investigación y Tecnología Agraria y Alimentaria, (INIA), grant number E-RTA2017-00003-C02-01, and Comunidad Autónoma de Madrid, grant number S2018/BAA-4370-ZOOVIR (PLATESA2-CM).

**Conflicts of Interest:** The author declares no conflict of interest. The funders had no role in the manuscript.

## References

1. Jenner, E. *An Inquiry Into the Causes and Effects of the Variolae Vaccinae, a Disease Discovered in Some of the Western Counties of England, Particularly Gloucestershire, and Known by the Name of the Cow Pox*; Sampson Low: London, UK, 1798.
2. Plotkin, S.A.; Mahmoud, A.A.; Farrar, J. Establishing a Global Vaccine-Development Fund. *N. Engl. J. Med.* **2015**, *373*, 297–300. [CrossRef] [PubMed]
3. Martín-Acebes, M.A.; Saiz, J.-C. West Nile virus: A re-emerging pathogen revisited. *World J. Virol.* **2012**, *1*, 51–70. [CrossRef] [PubMed]
4. Komar, N.; Langevin, S.; Hinten, S.; Nemeth, N.; Edwards, E.; Hettler, D.; Davis, B.; Bowen, R.; Bunning, M. Experimental Infection of North American Birds with the New York 1999 Strain of West Nile Virus. *Emerg. Infect. Dis.* **2003**, *9*, 311–322. [CrossRef]
5. Sejvar, J.J. Clinical Manifestations and Outcomes of West Nile Virus Infection. *Viruses* **2014**, *6*, 606–623. [CrossRef]
6. Bai, F.; Thompson, E.A.; Vig, P.J.S.; Leis, A.A. Current Understanding of West Nile Virus Clinical Manifestations, Immune Responses, Neuroinvasion, and Immunotherapeutic Implications. *Pathogens* **2019**, *8*, 193. [CrossRef]
7. Rizzoli, A.; Jiménez-Clavero, M.A.; Barzon, L.; Cordioli, P.; Figuerola, J.; Koraka, P.; Martina, B.; Moreno, A.; Nowotny, N.; Pardigon, N.; et al. The challenge of West Nile virus in Europe: Knowledge gaps and research priorities. *Eurosurveillance* **2015**, *20*. [CrossRef]
8. Bakonyi, T.; Ivanics, É.; Erdélyi, K.; Ursu, K.; Ferenczi, E.; Weissenböck, H.; Nowotny, N. Lineage 1 and 2 Strains of Encephalitic West Nile Virus, Central Europe. *Emerg. Infect. Dis.* **2006**, *12*, 618–623. [CrossRef]
9. Ladbury, G.A.F.; Gavana, M.; Danis, K.; Papa, A.; Papamichail, D.; Mourelatos, S.; Gewehr, S.; Theocharopoulos, G.; Bonovas, S.; Benos, A.; et al. Population Seroprevalence Study after a West Nile Virus Lineage 2 Epidemic, Greece, 2010. *PLoS ONE* **2013**, *8*, e80432. [CrossRef]
10. Ziegler, U.; Angenvoort, J.; Fischer, D.; Fast, C.; Eiden, M.; Rodriguez, A.V.; Revilla-Fernández, S.; Nowotny, N.; De La Fuente, J.G.; Lierz, M.; et al. Pathogenesis of West Nile virus lineage 1 and 2 in experimentally infected large falcons. *Vet. Microbiol.* **2013**, *161*, 263–273. [CrossRef]
11. De Oya, N.J.; Camacho, M.-C.; Blázquez, A.-B.; Lima-Barbero, J.-F.; Saiz, J.-C.; Höfle, U.; Escribano-Romero, E. High susceptibility of magpie (Pica pica) to experimental infection with lineage 1 and 2 West Nile virus. *PLoS Negl. Trop. Dis.* **2018**, *12*, e0006394. [CrossRef]
12. Venter, M.; Van Vuren, P.J.; Mentoor, J.; Paweska, J.; Williams, J.H. Inactivated West Nile Virus (WNV) vaccine, Duvaxyn WNV, protects against a highly neuroinvasive lineage 2 WNV strain in mice. *Vaccine* **2013**, *31*, 3856–3862. [CrossRef] [PubMed]

13. Merino-Ramos, T.; Blázquez, A.-B.; Escribano-Romero, E.; Cañas-Arranz, R.; Sobrino, F.; Saiz, J.-C.; Martín-Acebes, M.A. Protection of a Single Dose West Nile Virus Recombinant Subviral Particle Vaccine against Lineage 1 or 2 Strains and Analysis of the Cross-Reactivity with Usutu Virus. *PLoS ONE* **2014**, *9*, e108056. [CrossRef] [PubMed]
14. Minke, J.; Siger, L.; Cupillard, L.; Powers, B.; Bakonyi, T.; Boyum, S.; Nowotny, N.; Bowen, R. Protection provided by a recombinant ALVAC®-WNV vaccine expressing the prM/E genes of a lineage 1 strain of WNV against a virulent challenge with a lineage 2 strain. *Vaccine* **2011**, *29*, 4608–4612. [CrossRef] [PubMed]
15. Chaintoutis, S.C.; Diakakis, N.; Papanastassopoulou, M.; Banos, G.; Dovas, C. Evaluation of Cross-Protection of a Lineage 1 West Nile Virus Inactivated Vaccine against Natural Infections from a Virulent Lineage 2 Strain in Horses, under Field Conditions. *Clin. Vaccine Immunol.* **2015**, *22*, 1040–1049. [CrossRef] [PubMed]
16. Mukhopadhyay, S.; Kuhn, R.J.; Rossmann, M.G. A structural perspective of the flavivirus life cycle. *Nat. Rev. Genet.* **2005**, *3*, 13–22. [CrossRef]
17. Rathore, A.P.S.; John, A.L.S. Cross-Reactive Immunity Among Flaviviruses. *Front. Immunol.* **2020**, *11*, 334. [CrossRef]
18. Sinigaglia, A.; Peta, E.; Riccetti, S.; Barzon, L. New avenues for therapeutic discovery against West Nile virus. *Expert Opin. Drug Discov.* **2020**, *15*, 333–348. [CrossRef]
19. Krishnan, M.N.; Garcia-Blanco, M.A. Targeting Host Factors to Treat West Nile and Dengue Viral Infections. *Viruses* **2014**, *6*, 683–708. [CrossRef]
20. Angenvoort, J.; Brault, A.; Bowen, R.; Hermann, G.M. West Nile viral infection of equids. *Vet. Microbiol.* **2013**, *167*, 168–180. [CrossRef]
21. Byas, A.D.; Ebel, G.D. Comparative Pathology of West Nile Virus in Humans and Non-Human Animals. *Pathogens* **2020**, *9*, 48. [CrossRef]
22. Gould, L.H.; Fikrig, E. West Nile virus: A growing concern? *J. Clin. Investig.* **2004**, *113*, 1102–1107. [CrossRef] [PubMed]
23. Petersen, L.R.; Roehrig, J.T. Flavivirus DNA vaccines—Good science, uncertain future. *J. Infect. Dis.* **2007**, *196*, 1721–1723. [CrossRef] [PubMed]
24. Gardner, I.A.; Wong, S.J.; Ferraro, G.L.; Balasuriya, U.B.; Hullinger, P.J.; Wilson, W.D.; Shi, P.-Y.; MacLachlan, N.J. Incidence and effects of West Nile virus infection in vaccinated and unvaccinated horses in California. *Vet. Res.* **2007**, *38*, 109–116. [CrossRef]
25. De Oya, N.J.; Escribano-Romero, E.; Camacho, M.-C.; Blazquez, A.-B.; Martín-Acebes, M.A.; Höfle, U.; Saiz, J.-C. A Recombinant Subviral Particle-Based Vaccine Protects Magpie (Pica pica) Against West Nile Virus Infection. *Front. Microbiol.* **2019**, *10*, 1133. [CrossRef]
26. Gamino, V.; Höfle, U. Pathology and tissue tropism of natural West Nile virus infection in birds: A review. *Vet. Res.* **2013**, *44*, 39. [CrossRef]
27. De Oya, N.J.; Escribano-Romero, E.; Blázquez, A.-B.; Martín-Acebes, M.A.; Saiz, J.-C. Current Progress of Avian Vaccines Against West Nile Virus. *Vaccines* **2019**, *7*, 126. [CrossRef]
28. Chang, G.-J.J.; Davis, B.S.; Stringfield, C.; Lutz, C. Prospective immunization of the endangered California condors (*Gymnogyps californianus*) protects this species from lethal West Nile virus infection. *Vaccine* **2007**, *25*, 2325–2330. [CrossRef]
29. Samina, I.; Khinich, Y.; Simanov, M.; Malkinson, M. An inactivated West Nile virus vaccine for domestic geese-efficacy study and a summary of 4 years of field application. *Vaccine* **2005**, *23*, 4955–4958. [CrossRef] [PubMed]
30. Engle, M.J.; Diamond, M.S. Antibody Prophylaxis and Therapy against West Nile Virus Infection in Wild-Type and Immunodeficient Mice. *J. Virol.* **2003**, *77*, 12941–12949. [CrossRef]
31. Kaiser, J.A.; Barrett, A. Twenty Years of Progress Toward West Nile Virus Vaccine Development. *Viruses* **2019**, *11*, 823. [CrossRef]
32. Dayan, G.H.; Pugachev, K.V.; Bevilacqua, J.; Lang, J.; Monath, T.P. Preclinical and Clinical Development of a YFV 17 D-Based Chimeric Vaccine against West Nile Virus. *Viruses* **2013**, *5*, 3048–3070. [CrossRef] [PubMed]
33. Durbin, A.P.; Wright, P.F.; Cox, A.; Kagucia, W.; Elwood, D.; Henderson, S.; Wanionek, K.; Speicher, J.; Whitehead, S.S.; Pletnev, A.G. The live attenuated chimeric vaccine rWN/DEN4Δ30 is well-tolerated and immunogenic in healthy flavivirus-naïve adult volunteers. *Vaccine* **2013**, *31*, 5772–5777. [CrossRef]
34. Martín-Acebes, M.A.; Saiz, J.-C.; De Oya, N.J. Antibody-Dependent Enhancement and Zika: Real Threat or Phantom Menace? *Front. Cell. Infect. Microbiol.* **2018**, *8*, 44. [CrossRef]

35. Rothman, A.L. Immunity to dengue virus: A tale of original antigenic sin and tropical cytokine storms. *Nat. Rev. Immunol.* **2011**, *11*, 532–543. [CrossRef]
36. Kimura, T.; Sasaki, M.; Okumura, M.; Kim, E.; Sawa, H. Flavivirus Encephalitis. *Vet. Pathol.* **2010**, *47*, 806–818. [CrossRef]
37. Siirin, M.T.; Da Rosa, A.P.A.T.; Newman, P.; Weeks-Levy, C.; Coller, B.-A.; Xiao, S.-Y.; Lieberman, M.M.; Watts, D. Evaluation of the efficacy of a recombinant subunit West Nile vaccine in Syrian golden hamsters. *Am. J. Trop. Med. Hyg.* **2008**, *79*, 955–962. [CrossRef] [PubMed]
38. Konishi, E.; Yamaoka, M.; Win, K.-S.; Kurane, I.; Takada, K.; Mason, P.W. The Anamnestic Neutralizing Antibody Response Is Critical for Protection of Mice from Challenge following Vaccination with a Plasmid Encoding the Japanese Encephalitis Virus Premembrane and Envelope Genes. *J. Virol.* **1999**, *73*, 5527–5534. [CrossRef] [PubMed]
39. Beasley, D.W.C.; Li, L.; Suderman, M.T.; Guirakhoo, F.; Trent, D.W.; Monath, T.P.; Shope, R.E.; Barrett, A.D. Protection against Japanese encephalitis virus strains representing four genotypes by passive transfer of sera raised against ChimeriVax™—JE experimental vaccine. *Vaccine* **2004**, *22*, 3722–3726. [CrossRef] [PubMed]
40. Murray, K.O.; Mertens, E.; Desprès, P. West Nile virus and its emergence in the United States of America. *Vet. Res.* **2010**, *41*, 67. [CrossRef] [PubMed]
41. Zohrabian, A.; Hayes, E.B.; Petersen, L.R. Cost-effectiveness of West Nile Virus Vaccination. *Emerg. Infect. Dis.* **2006**, *12*, 375–380. [CrossRef]
42. Shankar, M.B.; Staples, J.E.; Meltzer, M.I.; Fischer, M. Cost effectiveness of a targeted age-based West Nile virus vaccination program. *Vaccine* **2017**, *35*, 3143–3151. [CrossRef]
43. West Nile Virus in Europe in 2020—Human Cases Compared to Previous Seasons. Available online: https://www.ecdc.europa.eu/en/publications-data/west-nile-virus-europe-2020-human-cases-compared-previous-seasons-updated-8 (accessed on 17 November 2020).


**Publisher's Note:** MDPI stays neutral with regard to jurisdictional claims in published maps and institutional affiliations.

MDPI
St. Alban-Anlage 66
4052 Basel
Switzerland
Tel. +41 61 683 77 34
Fax +41 61 302 89 18
www.mdpi.com

*Pathogens* Editorial Office
E-mail: pathogens@mdpi.com
www.mdpi.com/journal/pathogens

www.ingramcontent.com/pod-product-compliance
Lightning Source LLC
LaVergne TN
LVHW070600100526
838202LV00012B/518
* 9 7 8 3 0 3 6 5 7 2 4 0 6 *